HUG A TREE

And Other Things To Do Outdoors With Young Children

Robert E. Rockwell
Elizabeth A. Sherwood
Robert A. Williams

Illustrations by
Laurel J. Sweetman

Gryphon House, Inc.
Mt. Rainier, Maryland

Fourth printing, 1986
Third printing, 1985

ISBN O-87659-105-5

Published by Gryphon House, Inc., 3706 Otis Street,
Mt. Rainier, Maryland 20712.

Cover photo by Larry Wertz
Design: Cynthia Fowler
Typesetting: Lithocomp

To Our Children
Susan and Janet
Jennifer and Will
Jeff and Sissy
and to
All Children

May this book
bring you
joy

FOREWORD TO PARENTS AND TEACHERS

Children are active people, continually seeking to learn more about their world and how to function within it. This process involves curiosity, trust, and a desire to learn. Much of the current research in early childhood education suggests that children's basic attitudes toward life, their approach to new experiences, and their feelings about themselves and others are established in the first few years of life. It is safe to assume that their attitudes toward the environment are forming then as well. What better gift can we as parents and teachers give this precarious earth than a generation of children who have learned to know and love the natural world? Many of us admire a brilliant sunset, a breathtaking mountain vista, or a lovely array of spring flowers, and that is important. But it is even more important, in our opinion, to have people who, in addition to seeing the beauty, can value and care for all aspects of the environment. The world needs people who may not necessarily love snakes, but who appreciate the role they play in the environment. It needs people who understand the need for worms, dead trees, and marshes. This book is designed to help both children and the adults who are such an important part of their lives become understanding friends of our world.

Those who work with young children know that three year olds learn best through direct experience; they retain little of what is only talked about, but remember a surprising amount of what is touched, explored, and experienced. It is the adult's job to extend the children's initial explorations and to help them develop a framework for further learning. This can be done by providing the language which will give a means of talking and thinking about an occurrence. The framework grows as adults introduce children to activities which serve to broaden their initial experiences and to encourage more complex thinking. It should be stressed that the adult's role is that of a facilitator who supports the children's investigations rather than that of a leader who imposes a direction. This means that the adult is there to provide, as needed, ideas, materials, vocabulary, and safety. For example, making a tape recorder available and providing directions for its use, after a discussion of nature's noises, may open a whole realm of possibilities to children who have never used one. The activities in this book have been designed with this approach in mind. The activities serve to take the child from a simple observation, such as noticing a bird flying, to more complex concepts such as, identifying the bird as a blue jay, or grouping animals in the backyard into those that fly and those that don't. While the activities do not have to be used in sequence, the chapters do build on each other. The first two chapters contain suggestions to help children begin to appreciate their environment and to focus on what they see. These first activities should be used with children who are not familiar with the out-of-doors and who have not done much observing. The simplest activities begin each chapter and more complex ones follow. Within each activity, the "Want To Do More" section will probably require more skills and be more difficult than the initial section. The more experienced children can begin just about anywhere. Adults who know the children they are working with can choose appropriately.

CONTENTS

The age appropriateness given for each activity is an approximate, and is based upon our observations of many children who have tried them. Remember, each child has a particular backlog of experience which will guide any responses to an activity. The children and their reactions should be the final judges of whether the activity is appropriate for them or not.

HOW TO USE THIS BOOK

The sections are designed to give an adult, as briefly and simply as possible, all the information needed to successfully carry out the activities. The short introduction to each activity should provide enough information to help you decide if it will be appropriate for your children. The *2 & UP, 3 & UP, 4 & UP* and *5 & UP* boxes appearing at the beginning of each activity indicate the age group for which that activity is geared.

The "Things You Can Use" section lists materials needed for the activity. In most cases, these things are free or inexpensive. Other items can often be substituted. Read through the activity and use what is available; however, most schools or homes will have the suggested materials.

The "Words You Can Use" section contains vocabulary appropriate for each activity. Some words may sound too advanced for three year olds, at first reading, but keep in mind that the words are not to be memorized by the children. We do feel that exposure to them is important. All of us learn new words by hearing them used frequently in a meaningful context. Three year olds are no exception. You will be amazed at how quickly young children learn to use many of the terms. Those they don't learn will still be words with which they are familiar and could eventually use.

While learning the names of things is not the most important goal for young children, it does have value. Not only is knowing the name of a creature or plant fun, but it is a means of communicating about and focusing in on observations. One three year old we know spent a winter watching birds at a birdfeeder. He effortlessly learned the names of the frequent visitors. He enjoyed sharing this knowledge with his friends: he frequently made comments such as, "Today I saw the cardinals and the daddy downey woodpecker." Because of his regular observations, the first robin of spring came as a surprise. He queried "Who's that guy out there? I never knew his name before." It wasn't long before "robin" was a solid part of his vocabulary. No one taught him bird identification. People used the appropriate names in conversation and looked up those they didn't know. As a result, he could communicate accurately about what he had seen and could ask meaningful questions. All those involved would have been bored in a few short weeks with, "Look at that pretty bird."

The "What To Do" section consists of simple directions for the activity with the children. We recommend that you think the activity through from start to finish and that you ask yourself some questions. Do your children have the skills to do this activity or should they do some others first? (For example, children need to be able to sort things into groups before they can do some of the more complex classifying activities.) Where is the best place to do the activity? How many children should participate at one time? Is it something we will all enjoy?

The "Want To Do More" section provides suggestions for building on the initial activities. There are usually ideas at the same skill level and a few which are more elaborate. You may find that the "Want To Do More" activities are more suitable for your children than the main ones. Always use what works best for the children. Reading the "Want To Do More" section before beginning will supply numerous possibilities for extending or adapting the experience while it is in progress rather than waiting for another opportunity.

EXPANDING THE OUTDOOR EXPERIENCE

One of the most effective ways to help children consolidate and retain information is to record that information in some way. For the very young, this most often means using real objects, pictures, and their own words.

The examples which follow have been divided into two sections, those which involve descriptive language and those which record data.

Descriptive Language

Outdoor activities provide a wealth of experience which the children want to communicate. Essential language skills begin when children realize that writing is the spoken word put down on paper. Here are some ways to start:

Getting Kids to Talk - A List of Ideas

1. Vary the setting. Many children who won't talk in a large group will speak freely in a small group. At home, encourage even little ones to talk at the dinner table. An outdoor event gives them something important to tell others. Don't forget one-to-one conversations, especially to bring the shy ones out. A suggestion such as "Go tell Sybil about the feather you found," encourages both information and conversation.

2. Be a good model. Share your own discoveries with enthusiasm, and be a good listener as well.

3. Try talking games. Use riddles, guessing games, or hide and seek with verbal clues. Objects collected on walks, animals observed, and places visited can all be described. These games help children learn to be careful listeners while also increasing their vocabulary.

4. Reminisce about activities. How many details can you recall? This not only helps develop memory, it says to the children, "What we did was worth remembering."

Getting the Talk Written Down

1. The simplest approach is to have the children dictate a few brief statements about an experience. Write these sentences using the children's own words; these can be read on later occasions to recall an event. Simple sketches or collected objects may accompany the words.

Example: I saw a millipede. He rolled up and I poked him. I didn't bother him so he got flat and he walked away and he went home to his mama.

2. Another approach is for the adult to provide a structure to follow. Often a repetitive pattern can lend a poetic quality to the writing. The following example was written with a group of three and four year olds.

We hear the wind in the leaves.
We hear the water go by.
We hear the birds.
We hear the wind in the grass.
We heard that airplane make noise.

While this is certainly not a masterpiece, it led to a discussion of man-made and natural sounds. Such writing can focus on anything, and can be as long or short as desired.

Examples: Green is clover.
Green is grass.
Green is a dandelion stem or a dandelion leaf.

What do you see?
Jay sees a robin.
What do you see?
Gerald sees a big tree.
What do you see?
Molly sees a bunch of grass.
What do you see?
Tia sees a lots a pile of rocks.

Preserving the children's own language, as was clearly done in the last example, is important. It demonstrates to the children that their thoughts are accepted as valuable just the way they are, and that the written word is real language. Correct sentence structure can be emphasized later.

3. Art work and diagrams also provide an opportunity for the use of written language in the form of labels or captions. Again, the children's own wording should be used. Let us mention here that it is much more tactful to say to a child, "Will you tell me about your work?" than to say, "What is it?" Keep in mind that much of young children's art is done for sensory pleasure. They enjoy mixing colors, moving their arms freely, sliding paints across paper, and pasting one item on another. The completed work may not "be" anything. A child's caption may be "I used the whole lid of paste. It glued every leaf on", rather than "It's a tree." It is also fun to caption a photograph of an outing or of a familiar place.

Recording Data

Children can organize and react to objects and events encountered outdoors if the information is recorded in some way. This information we call data, and it can be associated in a logical, meaningful and often mathematical order. The primary function of the following activities is to record in a visual display the children's experiences.

Activity Charts

An activity chart is used to record the activities of the people in a group. It increases awareness of the activities of others and is a permanent record of an experience. Kids' names are the only words required, although a few additional defining words may be used. Symbols, such as arrows or drawings, along with real objects, create a record which three year olds can "read" once they are familiar with the process.

Examples: Following a leaf collecting walk

(Found)

Jacob - - - - - - - - ->

Maggie - - - - - - - - ->

The discussion described in "What Makes a Perfect Day?" on page 38 could be summarized with this method.

(Likes to)

Troy - - - - - - -> (fly kites)

Kari - - - - - - - -> (camp)

Simple sketches are all that are required: those with no artistic talent need not despair. Each chart should be titled, dated, then read by the group. This is best suited to small groups, or a chosen few from a larger group. Using more than five or six children changes the activity from fun to tedious.

Real Graphs

Object graphs use the actual objects in making comparisons. The simplest one compares the quantity of two things.

Example: Did we find more acorns or walnuts?

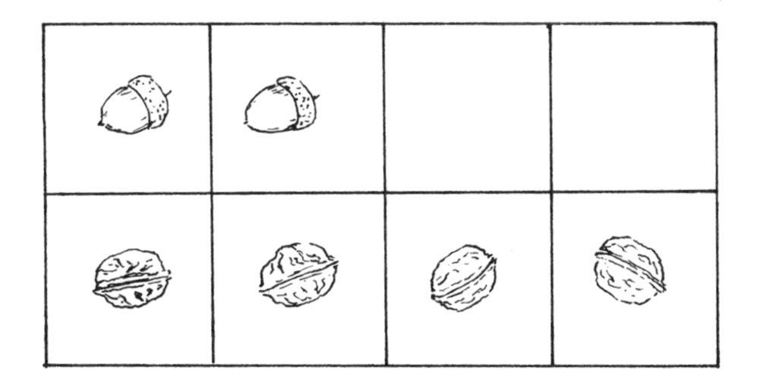

Notice that we have a title, so the goal is clear, and a baseline, or starting point, so accurate comparisons can be made. For example, to determine which yielded the greater harvest, pumpkins or squash, two large columns of squares can be marked off on butcher paper. One fruit is placed in each square. Children don't need to be able to count to know which group "won". A permanent graph for large objects can be drawn on a plastic window shade or other heavy plastic. Make two columns on one side and four on the other to increase versatility.

Picture Graphs

From object graphs, children can usually move quickly to using pictures to represent the real objects. Again, format is important so that the results are accurate. Pictures used should be uniform in size or fit into uniform squares. The paper on which they are placed should be titled and a baseline established.

Example: What kind of leaves did we find?

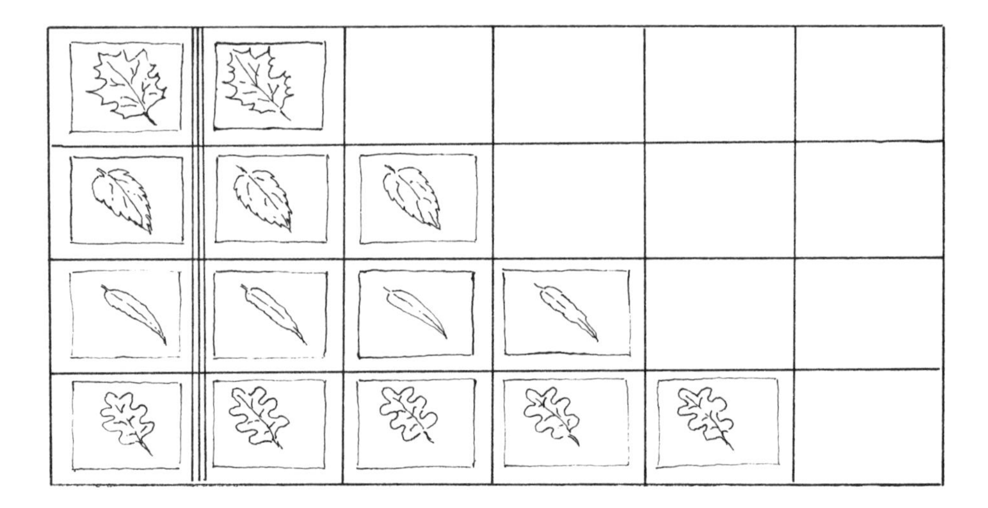

The reason for graphing can be made obvious to children by going through the following activity. Choose a simple question such as "Did most people wear boots or shoes today?" Give each child a 3x5 card with a picture of a boot or shoe on it, depending on what they wore (they can drawn their own if possible). Have the children place their cards on the titled paper anywhere they like. Ask for an answer to the question of boots or shoes. Assuming there is a fairly even distribution (check it out before you start!), it will be difficult to answer without careful counting. Repeat the activity using the proper graphing structures. The value of organizing information should be obvious to everyone.

Bar Graphs

When the children are comfortable with picture graphs, they can begin to use the more conventional bar graphs. An excellent way to begin is to use a two column graph. Mark off the squares and ask each child to color a square in the appropriate column. Using a one-to-one correspondence is the key to understanding graphing at this age.

Example: What do you think is the best spot for exploring; the woods or the creek?

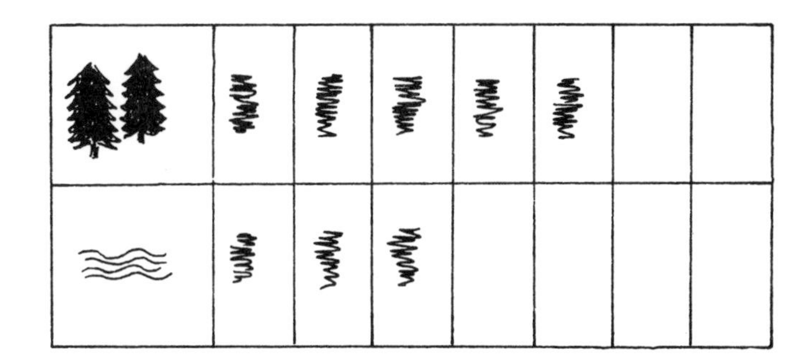

This method can also be used for tallies. Send the children out to find the colors of spring (page 93). Construct a chart and color in the appropriate squares.

RED	ORANGE	YELLOW	GREEN	BLUE	VIOLET	BROWN	BLACK	

For small objects, egg cartons or the bottoms of milk cartons are ideal. These can be stapled together in whatever form is needed. As stated earlier, these are ways of struc-

turing the graphing process to make it both easy and accurate for young children. In this stage, counting is purely optional. What is more important is a visual means of making comparisons.

Tally Cards

A simple way to do tallies with very young children is to place one child in charge of each category. For example, on a bird walk give one child a card with bluejay at the top, another a robin card, and so on. Whenever anyone observes a bluejay, they notify the bluejay person, and he or she records the observation with a mark on the card. Robins and other kinds are recorded similarly. At the end of the walk the cards can be compared, or a chart constructed.

Graphing and chart making should be repeated in many situations. Increase the complexity gradually, until the children are able to work almost alone. Three year olds derive real satisfaction from creating a record that they can refer to later. A chart or graph made in November and referred to in March helps children realize the importance of keeping a record of personal experiences. Such record keeping shows the children that their work, like that of adults, is important enough to be preserved. In addition, it demonstrates that all those marks adults make on paper have permanent meaning. These activities also provide a means of introducing children to reading and math concepts in a relaxed, non-threatening way.

OUTDOOR EDUCATION ISN'T ALWAYS IN THE WOODS

One of the assumptions we make in assembling this book is that you do not need a deep dark woods or a meadow in order to experience the natural world. In fact, any area that contains a sprinkling of plant life can serve as an outdoor laboratory. However, it would be difficult to conduct an entire environmental program in such a restricted environment as a paved school yard. You should plan to look within a three to four block area for extras.

A vacant lot, whether overgrown with weeds or covered with a tangle of shrubs, provides an area for study. The edges of sidewalks, planters, and small city pocket parks all provide a portion of the resources needed to develop an understanding of how the environment works.

Once the children have begun to develop a sense of how they interact with nature in their own environment, then you can expand their horizons to larger and more complex natural settings. This is the time for trips to larger parks or nature preserves and, finally, out to the open countryside, rivers or beaches.

Those of you using the book who are lucky enough to have wild areas nearby have a wealth of material. Yet even in the concrete of the city the wild exists, though it is not as evident. The real value of our wild lands cannot be appreciated by our children unless they see the value of living things in *all* environments.

The activities described in this book will give you a format and a direction for developing your own environmental curriculum. Remember that every setting will change the results. The experiences you have on the beach will differ greatly from those in a city park. To draw contrasts and to emphasize patterns, some activities should be repeated in different settings. In this way, the connections and the similarities that we adults have learned to see between differing environments can be made accessible to children.

HOW TO ORGANIZE FOR AN OUTDOOR EXPERIENCE

The authors have developed a checklist to help you move step by step through the planning of your excursions. All of the steps listed are essential, although some may be done only once if walking is the means of transportation, or if your activities are conducted regularly.

Obviously, parents and teachers will have different approaches to planning outdoor experiences. The following ideas are directed primarily to teachers; however, parents should consider some of the suggestions, omitting those which pertain specifically to schools. Building on ideas and prior experiences is just as important at home as it is at school.

A teacher's first consideration should be whether this trip has value to the overall curriculum. Because many school systems have reduced funding for extra activities, the teachers must gain the support of administrators and boards if field trips are to remain part of the school's instructional tools. If you have developed a continuous curriculum with environmental experiences as a part, you have gone a long way in establishing a sound basis for the existence of these experiences.

Pretrip Planning

One of the most important parts of an outdoor experience is the planning. If this is thoroughly done then most details will fall into place. The spontaneous walk on a spring day should not be unplanned. Even if the field experience is for the pure joy of seeing the world outside, a simple plan should exist and you should have thought out the trip.

After your planning (see checklist), conduct any pretrip experience with the children. If your trips outdoors are frequent, so that the children are involved in an ongoing process, this preparation may be very short or nonexistent. The goal of the pretrip exercise is to prepare the children to get the most out of the experience. You should discuss the trip, list any objectives and review information or background that is needed. You set the tone and talk about the information you hope to obtain.

You should also make a list and notify parents of materials or clothing needed for the trip (see sample letter). Some things to include in pretrip planning are:

1. Clothing
2. Time, bathrooms, snacks and water (pre-walk your trip taking these into consideration)

3. Accidents—carry first aid for scratches and bee stings, but plan what to do in case of something serious.

4. Rain plans.

5. Alternatives in case of cancellation.

6. Rules, i.e. partners, aides, and lines.

7. Children's participation in planning.

The Trip

Conduct the trip or activity as planned, but always allow for the unexpected. Those special happenings can change the focus of a lesson or greatly add to its breadth. Some discoveries that come along can be left to a later exploration. Others must be followed up now, at the expense of your planning.

It is especially important during early lessons to develop a sense of organization and continuity so that later experiences require less teacher direction and more teacher-child cooperation. Along with good preparation, a real involvement of the child in the total experience is a must.

Follow-Up

Repetition is a key to the young child's memory development. Follow-up after the trip becomes a must. If a child is to develop an understanding of and love for the environment, repeated exposure to the processes of the environment is essential.

For follow-up, it is a good idea to keep a log, draw or tape record the trip. Tell the story of the trip and have each child imagine the time when the experience was in full progress. It is also helpful to select books that support the trip. These can also be used in pretrip planning. You may plan an art experience, a storytelling session, or recall the trip through show and tell. Evaluate the experience from both your own and the child's perspective. As you begin planning the next excursion, see how this one can build on the last one and on the total experiences of the child. Keep your total curriculum in mind.

A Checklist of Field Trip Priorities

1. Determine the reason for going.

2. Select the site.

3. Discuss the site permission and availability with proper persons. Set tentative dates.

4. Estimate trip costs, if any, and determine who is to pay them.

5. Obtain administrative and/or Board approval.
6. Select and contact resource people for the trip or site. Set time and date. Confirm reservation on day you go.
7. Visit the site and plan travel route.
8. Locate aides, volunteers or chaperones.
9. Obtain parental permission. Include special clothing needs, funds needed, and other specific plans on this permission slip. (See sample permission slip page)
10. Conduct a planning session with all involved.
11. Ready equipment and supplies.
12. Establish trip plans and rules with staff and children. Plan for medical emergencies.
13. Conduct any pretrip exercise with children.
14. Take the field trip.
15. Conduct post trip exercises, including evaluation of the trip and commentary for using this activity again.

Suggested School Forms

Dear ____________________________,

The ____________________________ Child Care Center is preparing to conduct a series of outdoor experiences of which __________________________ will be a part. We will be walking to the area around the center and will be conducting a field experience every other day, weather permitting.

During the walks your child will be involved in a series of outdoor experiences that are designed to help foster environmental awareness and concern. These activities were developed for children and are available in a book entitled ____________________________. This book can be purchased in the ______________________ book store, if you wish to continue the activities at home.

Because your child may be leaving the Center premises during these experiences, we at the Center will need to update our special health form and other personal information. Please fill out the form at the end of this letter and return it, signed.

As your child's experiences become more complex we will be taking longer walking trips and will finally embark on all day trips to ____________________ State Park. These trips will require much more planning and there may be some costs with which we hope you can help. We also hope you might help on these longer trips by serving as a parent volunteer.

Sincerely,

____________________________, Director

- -

Student's name ____

Phone numbers to call in case of emergency (1) ______________ (2) ______________

I grant permission for my child to be involved in your environmental program.

________________________________ ________________________________

date signature of parent

Does your child have any physical limitations? ( )Yes ( )No

Comment on type ____

Does your child have any allergies? ( )Bee stings ( )Poison ivy ( )Other

Comment on type ____

Is your child under a doctor's care at this date? ( )Yes ( )No

Comment ____

Is your child taking any type of medicine? ( )Yes ( )No

Comment ____

Has your child had a tetanus shot? ( )Yes ( )No

I will serve as a volunteer, call me. ______________________________________

YOU'RE GOING TO NEED HELP

This book provides a beginning for environmental education. Where do you go for ideas and resources when you are ready to do more? If you live in an urban area, you have access to museums and zoos. These places are staffed with people who can give you answers. Remember, you will have to ask! Many questions can be answered over the telephone. A librarian can often help find exactly the right picture of a frog or a specific bird. Children's librarians know their selections well and will gladly assist you in finding appropriate children's books on any subject. They can also suggest films, video tapes, records, posters or other resources.

Whether you live in the country or the city, try the cooperative extension service (U.S. Department of Agriculture) in your county. Extension publications provide a wealth of background information and ideas for doing things out of doors (e.g., indoor and outdoor gardening). If you prefer to write, send your letters of request to the state cooperative extension service at your state land grant university. Public Health Services, state or local, are good places to get safety, health and nutrition aids and general information, too.

Don't forget the field offices of other government services. Forest Service (U.S. Department of Agriculture), Fish and Wildlife Service (U.S. Department of Interior), Soil Conservation Service, or Department of Natural Resources are all good contacts. Public education is a major part of the job of these government services. They realize that the natural resources they manage and protect will fare much better at the hands of an informed public. Usually these staff people are glad to help.

Private groups can also provide both information and possible field trip sites. A local businessman may own a farm which has a wooded area or pond to explore. You won't see it if you don't ask! Private individuals are often eager to share their resources with families or with preschool groups.

Some of the best sources can be your friends. Adults learn from each other. Sharing your activities and your interest in outdoor education can uncover some new resources for everyone. By talking to each other we can accumulate a wealth of wisdom that can be shared by young and old alike.

Books for Parents and Teachers

Abruscato, J., & Hassard, J. *The Whole Cosmos: Catalog of Science Activities.* Salt Lake City: Goodyear, 1977. A collection of over 250 activities in the life sciences, the earth sciences, the physicial sciences, the aerospace sciences, and speculative fiction.

Althouse, R., & Main, C. *Science Experiences for Young Children.* New York: Teacher's College, 1975. Available in a boxed set. Ten booklets on the biological and physical sciences that provide a variety of activities arranged in sequential order.

Brandwein, Paul F., and Elizabeth K. Cooper. *Concepts in Science.* New York: Harcourt, Brace & World, 1980. An illustrated book without text, showing scenes such as children playing with magnets and levers, chicks hatching, etc. Sections cover investigating matter, investigating force, investigating plants and animals.

Cobb, V. *Science Experiments You Can Eat.* Philadelphia: Harper & Row, 1972. Using the kitchen as a laboratory, this book provides a series of experiments and explanations for the changes that take place in food processing and preparation.

Harlan, J. *Science Experiences for the Early Childhood Years.* Columbus, Ohio: Merrill, 1979. Ideas for designing small-group activities for preschool and elementary school children to reveal scientific concepts.

Hillcourt, W. *The New Field Book of Nature Activities and Hobbies.* New York: Putnam, 1978. Close to 1,000 activities designed to foster children's appreciation of nature.

Hounshell, Paul B., and Ira Tollinger. *Games for the Science Classroom: An Annotated Bibliography.* Washington, D.C.: National Science Teacher Assoc., 1977. One hundred thirty instructional games and simulations for science teaching, organized under the categories of biological science games, physicial science games, earth and space science games and general science games.

Gale, Frank C., and Clarice W. Gale. *Experiences with Plants for Young Children.* Palo Alto, CA: Pacific Books, 1975. A resource book for use primarily with four-and five-year-olds. Two categories of experiences are provided. In the first, entitled "Exploring," the observational power of the child is sharpened, but no comparison of materials is made. In the second, "Exploring, Comparing, and Seeing Relationships," the child is asked to make comparisons between two or more objects, events, or ideas.

Project Learning Tree, Activity Guide for Grades K through 6. The American Forest Institute, Inc.: 1619 Massachusetts Avenue, N.W., Washington, D.C. 20036, 1977. An excellent resource for environmental education curriculum. Must be adapted for preschool level. To obtain information and free materials contact the American Forestry Association.

Rieger, Edythe. *Science Adventures in Children's Play.* New York: The Play Schools Association, 1968. This is a booklet written for teachers of elementary school science, but the preschool teacher will find it useful as a resource for ideas to incorporate into his or her curriculum.

Schmidt, V., & Rockcastle, V. *Teaching Science With Everyday Things.* New York: McGraw-Hill, 1968. Emphasizes activities that do not require special or costly equipment. Covers such areas as air and weather, sun, moon, and stars, forces and motions.

Skelsey, A., & Huckaby, G. *Growing Up Green.* New York: Workman, 1973. Source of information about plants and gardening. Suggests appropriate activities for preschoolers.

Tarrow, Norma B. and Sarah W. Landsteen. *Activities and Resources: For Guiding Young Children's Learning.* New York: McGraw-Hill, 1981. An excellent resource for science activities for 3, 4, & 5 year olds.

Vivian, Charles. *Science Experiments & Amusements for Children.* New York: Dover, 1967. This booklet contains seventy-three simple experiments which young children can do.

Science and Nature Magazines

Children's Playmate
Review Publishing Co.
1100 Waterway Blvd.
Indianapolis, IN 46202

Simple articles, illustrations, and experiments on nature, science, and general topics.

Curious Naturalist
Massachusetts Audubon Society
S. Great Road
Lincoln, MA 01773

Suggestions for simple outdoor activities.

Journal of Outdoor Education (FREE)
Northern Illinois University
P.O. Box 299
Oregon, IL 61061

Provides parents and teachers with suggestions for various outdoor education activities.

Owl
The Young Naturalist Foundation
59 Front St. East
Toronto, Ontario
Canada M5E1B3

Information on an abundance of different activities for the outdoors.

Ranger Rick's Nature Magazine
National Wildlife Federation
1412 16th St., N.W.
Washington, D.C. 20036

Provides activities and information to help children enjoy nature.

Chickadee
The Young Naturalist Foundation
59 Front St. East
Toronto, Ontario
Canada M5E1B3

Excellent for 3–5 year olds.

Your Big Backyard
National Wildlife Federation
1412 16th St., N.W.
Washington, D.C. 20036

Perfect for kids 3–5. Wildlife photos, stories, puzzles, games and poems.

Books For Children

Baer, Edith, *The Wonder of Hands.* New York: Parent's Magazine Press, 1970. The text and sensitive photography portray the many ways hands communicate: hands can heal, plant a seed, finger-paint, wave goodby.

Bartlett, Margaret Farrington, *The Clean Brook.* New York: Harper & Row, 1960. The story of the changing life of a brook.

Bonforte, Lisa, *Farm Animals.* New York: Random House, 1981. A simple introduction for the preschooler to well known farm animals.

Brandwein, P. F., & Cooper, E. K., *Concepts in Science.* New York: Harcourt, Brace and World, 1980. Illustrated text; vocabulary at end of book; many science areas covered.

Branley, Franklyn M., *Sunshine Makes The Seasons.* New York: Crowell, 1974. Describes how sunshine and the tilt of the earth axis are responsible for the changing seasons.

__________, *What Makes Day and Night?* New York: Harper & Row, 1961. A simple, detailed explanation describing the mechanics of the earth's rotation.

Bruna, Dick, *My Shirt Is White.* New York: Methuen, 1975. The bright colors of the illustrations make it an excellent early book about colors.

Carle, Eric, *My Very First Book of Shapes.* New York: Harper & Row, 1974. Lets the child match shapes on the top half of the page with a like shape on the bottom half of the page.

________, *The Very Hungry Caterpillar.* New York: World, 1971. A book to use in showing stages of growth in a caterpillar.

Chen, Tony, *Wild Animals.* New York: Random House, 1981. A simple introduction for the preschooler to some well known wild animals.

Collier, Ethel, *Who Goes There In My Garden?* New York: Addison Wesley, 1963. A good book to read to children before planting a garden.

Cornelius, Carol, *Robin's Land.* Elgin: Child's World, 1978. Follows the life of a robin family from their arrival to their summer home in early spring to their southern migration in the late fall.

Crews, Donald, *Light.* New York: Greenwillow, 1981. A graphic presentation of kinds of light—daylight, starlight, lightning, electric signs, etc.

Dabcovich, Lydia, *Follow The River.* New York: Dutton, 1980. Follows a stream from its home in the mountains through the countryside where it becomes a river and eventually flows into the ocean.

Davis, Jane, *Why Does The Tiger Have Stripes?* Elgin: Child's World, 1978. Rhyming answers to questions regarding the protective devices, particularly color, of such animals as the tiger, frog, polar bear and chameleon.

Engdahl, Sylvia, *Our World Is Earth.* New York: Atheneum, 1979. Introduction to the earth, sun, solar system and outer space.

Fisher, Aileen, *Anybody Home?* New York: Crowell, 1980. Describes in rhymed text the homes of a variety of animals.

Florian, Douglas, *A Bird Can Fly.* New York: Greenwillow Books, 1980. Describes activities that a bird, beaver, ant, tortoise, monkey, camel, and a fish can and cannot do.

Fuchshuber, Annegart, *From Dinosaurs to Fossils.* Minneapolis: Carolrhoda, 1981. Discusses dinosaurs that range in size from that of a chicken to five times that of an elephant.

Fujikawa, Gyo, *Let's Grow A Garden.* New York: Grosset and Dunlap, 1978. Bright, clear, full color drawings show a multi-cultural group of very young children planting and growing a garden.

Gans, Roma, *It's Nesting Time.* New York: Thomas Y. Crowell, 1964. A book designed to teach young children to observe and respect the nests of various birds.

Garelick, May, *Where Does The Butterfly Go When It Rains?* New York: Addison Wesley, 1961. Story concerns the "mystery" of where the butterfly goes when it rains. Encourages the child to look and discover the answers for himself. Blue-hued illustrations give the impression of rain.

George, Jean Graighead, *The Wounded Wolf.* New York: Harper & Row, 1978. As hungry animals close in on an injured wolf, hoping to feed on him after death, help arrives.

Goldreich, Gloria, and Esther Goldreich, *What Can She Be? A Veterinarian.* New York: Lothrop, Lee & Shepart, 1972. Beautiful photographs showing a typical day in the life of a woman vet caring for injured and sick puppies, cats, and bunnies.

Goldstein-Jackson, K., et al., *Experiments With Everyday Objects.* Englewood Cliffs, N.J.: Prentice Hall, 1978. Over seventy easy experiments for kids.

Hoban, Tana, *Circles, Triangles and Squares.* New York: MacMillan, 1974. Blowing bubbles, spinning a loop, building card houses or grating cheese, you can find a myriad of shapes.

____________, *Count and See.* New York: MacMillan, 1972. Clear photographs of familiar objects are easily recognized and fun to count.

____________, *Is It Red? Is It Yellow? Is It Blue?* New York: Scholastic, 1978. Clear color photographs which explore the identification of color.

Jessel, Camilla, *The Puppy Book.* New York: Methuen, 1980. Text and illustrations follow a retriever as she gives birth to and cares for her puppies.

Keats, Ezra Jack, *The Snowy Day.* New York: Viking Press, 1962. The story of a young boy's delight during his first snowfall.

Krauss, Ruth, *The Carrot Seed.* New York: Harper & Row, 1945. Illustrated. A little boy is convinced that his carrot seed will grow in spite of his family's doubts.

Kunhardt, Dorothy, *Pat the Bunny.* Calif. Western Publishing, 1962. Paul and Judy find something on each page to touch, smell and see. A pleasant sensory experience for very young children.

Kuskin, Karla, *A Space Story.* New York: Harper & Row, 1978. Takes young readers on a journey through space from a small boy's bedroom through our solar system and beyond. A fiction tale and a factual introduction to the sun and planets.

Leaf, Munro, *Who Cares? I Do.* New York: J. B. Lippincott, 1971. Cartoon-like figures help the child to see what can be done to protect the environment.

Lexau, Joan, *The Spider Makes A Web.* New York: Hastings House, 1979. Discusses, for beginning readers, the ways of the spider: an ordinary garden creature who is not so ordinary after all.

Linsenmaier, Walter, *Wonders of Nature.* New York: Random House, 1980. An exquisitely illustrated early science book filled with remarkable facts about animal habits and behavior.

Lionni, Leo, *Fish is Fish.* New York: Pantheon Books, 1970. A good book to use with tadpoles and fish in a science project.

McMillan, Bruce, *Apples, How They Grow.* Boston: Houghton Mifflin, 1979. Describes how apples grow, from dormant bud to ripe fruit.

Mari, Iela, and Enzo Mari, *The Apple and The Moth.* New York: Pantheon, 1970. Lovely picture book about the metamorphosis of a moth and its stages—egg on leaf, caterpillar, cocoon, moth. A book without words which tells its own story.

__________, *The Chicken and The Egg.* New York: Pantheon Books, 1970. A hen lays an egg and the following sequence of pictures shows the development of the chick until it finally hatches.

McGee, Myra, *Willie's Garden.* Emmaus, PA: Rodale Press, 1977. A little boy plants one plant at a time and soon finds himself with a big garden.

Reidel, Marlene, *From Egg to Butterfly.* Minneapolis: Carolrhoda, 1981. Describes the metamorphosis of a butterfly through its stages of egg, caterpillar, pupa, and finally butterfly.

__________, *From Egg to Bird.* Minneapolis: Carolrhoda, 1981. Describes the development of a bird from the time the egg is produced until the chick acquires adult plumage.

__________, *From Ice to Rain.* Minneapolis: Carolrhoda, 1981. Describes the cycle in which ice on the pond melts into water, which in turn evaporates into water vapor, which collects into clouds, which produces rain and snow.

Scarry, Richard, *Richard Scarry's Coloring Activity Book.* New York: Random House, 1974. Mr. Paint Pig introduces young children to colors in this brightly colored book.

Schlein, Mariam, *Shapes.* New York: Addison Wesley, 1952. Helps children learn to classify things by shape, using familiar objects.

Showers, Paul, *Find Out By Touching.* New York: Harper & Row, 1961. Helps the child to learn how his sense of touch can tell him many things about the world around him. Through touch, the child learns the concepts of hard, soft, smooth, rough, cold, warm, etc.

Snoopy's Fact and Fun Book About Nature. New York: Random House, 1980. Another Snoopy book with facts and fun about nature.

Snoopy's Fact and Fun Book About Seasons. New York: Random House, 1980. A small book which combines simple facts and humor through the courtesy of Snoopy and his friends.

Stone, A. Harris, *The Last Free Bird.* Englewood Cliffs, NJ: Prentice-Hall, 1967. Beautiful illustrations enhance text which tells of man's destruction of the nesting and feeding places of birds.

Tresselt, Alvin, *The Dead Tree.* New York: School Book Service, 1972. Presents the nature cycle of life in the forest where even a dead tree serves to enhance new growth.

Miller, Edna, *Jumping Bean.* Englewood Cliffs, N.J.: Prentice-Hall, 1979. Many animals wonder at the bean that jumps. In time its secret is revealed.

Mitgutsch, Ali, *From Seed to Pear.* Minneapolis: Carolrhoda, 1981. Describes the cycle of a pear seed which, when planted, produces a fruit-bearing tree and a supply of new seeds.

Moncure, Jane Belk, *Magic Monsters Learn About Space.* Illustrated by Patricia McMahon Boman, Elgin, IL: Child's World, 1980. The magic monsters visit and study about the planets.

Nockels, David, *Animal Builders.* New York: Dial, 1981. Simple yet informative text combined with pop-up illustrations depict animals in motion—running, jumping, swimming and flying.

__________, *Animal Marvels.* New York: Dial, 1981. Another pop-up book of the same type.

Pine, Tillie S. and Levine, Joseph, *Measurements and How We Use Them.* New York: McGraw-Hill, 1977. Projects and examples introduce various measuring devices and how to use them.

Provensen, Martin and Alice, *A Book of Seasons.* New York: Random House, 1978. Spring, Summer, Fall and Winter are all illustrated in full color with the activities common to each season.

Toye, Elliam, *The Fire Stealer.* New York: Oxford, 1979. Retelling of a well-known Canadian Indian legend with unique collage illustrations concerning the introduction of fire and its many uses.

Udry, Janice May, *A Tree Is Nice.* New York: Harper & Row, 1956. A great book showing why trees are nice to have around. You can climb on them, eat their fruit, hang a swing, picnic in their shade, and plant your very own.

Wildsmith, Brian, *Animal Homes.* New York: Oxford University, 1980. Discusses where animals live, why they live there, what their homes are like and how they make them.

Zallinger, Peter, *Prehistoric Animals.* New York: Random House, 1978. Drawings and brief text describe a variety of prehistoric animals.

Scrounging Places and Things to Look For

Shoe stores (for boxes)
Ice Cream parlors (for storage containers)
Photography studios (canisters, empty film spools, chemical bottles)
Telephone company (colored wire)
Automobile repair shops (old magnets and motor parts)
Drug store (eyedroppers, old vials, bottles)
Nurseries, plant stores (seeds, plants, planters, old pots)

Field Trip Ideas, Guest Speakers and General Information

Government conservation sites
Dams or reclamation projects
Telephone companies
Museums of natural history
Museums of science or industry
Hatcheries
Nurseries
Taxidermists
Game wardens
Forest rangers
Pollution control officers
Sierra Club, or local conservation groups
Botanical gardens
Zoos
Planetariums
Local parks departments
Fire departments
Utility companies
Doctors and dentists
Sailors and navigators
Geologists
Other scientists
Weather forecasters
Aquariums
Various outdoor sites (parks, forests, farms and lakes)
University faculty

Things To Use

Magnifiers and microscopes (those designed for preschool use)
Scales
Measuring sets (spoons and cups)
Magnets
Weather instruments (thermometer, barometer, rain gauge, weather vane)
Stethoscope
Anatomical models (for instance, animal skeletons, preserved birds, mounted insects)
Dinosaur models
Rocks
Sand
Fossils
Bones

Gardening tools
Seeds
Water
Live Plants
Terrariums
Animals
Cages
Aquariums
Butterfly nets

Places to Write

American Assn. for Health, Physical Education, and Recreation
1202 16th St., NW
Washington, DC 20036

American Fisheries Society
1040 Washington Bldg.
Washington, DC 20005

American Forest Products Industries, Inc.
1835 K St., NW
Washington, DC 20006

American Forestry Assn.
919 17th St., NW
Washington, DC 20006

Animal Welfare Institute
P.O. Box 3492
Grand Central Sta.
New York, NY 10017

Environmental Defense Fund
P.O. Drawer 740
Stony Brook, NY 11790

Friends of the Earth
30 East 42nd St.
New York, NY 10017

National Audubon Society
950 Third Avenue
New York, NY 10022

National Geographic Society
17th and M Sts., NW
Washington, DC 20036

National Parks Assn.
Washington, DC 20005

National Wildlife Federation
1412 16th St., NW
Washington, DC 20036

Division of Surveys and Field Services
Peabody College
Nashville, TN 37203

Sierra Club
1050 Mills Tower
San Francisco, CA 94104

U.S. Bureau of Mines
Washington, DC 20250

U.S. Chamber of Commerce
National Resources Dept.
161 H St., NW
Washington, DC 20036

U.S. Environmental Protection Agency
Consumer Information Center
Pueblo, CO 81003

U.S. Fish and Wildlife Service
Washington, DC 20250

U.S. Forest Service
Washington, DC 20250

U.S. Geological Survey
Washington, DC 20250

U.S. Indian Service
Washington, DC 20250

U.S. Naval Observatory
Washington, DC 20250

U.S. Office of Education
Washington, DC 20250

U.S. Soil Conservation Service
Washington, DC 20250

U.S. Weather Bureau
Washington, DC 20250

Wilderness Society
729 Fifteenth St., NW
Washington, DC 20005

LOVING YOUR ENVIRONMENT

Aesthetic & Affective Experiences

SPYGLASS TREASURE HUNT: CLOSE-UP AND FAR AWAY

Things You Can Use

paper rolls
(i.e. toilet paper, paper towel, foil, or wrapping paper rolls)
magnifiers
telescope
binoculars
microscope

Words You Can Use

focus
telescope
view
binoculars
field of vision
cylinder

Looking through a cylinder limits the scope of the eye and forces the observer to focus on specific rather than general aspects of the environment. In this activity, the children are encouraged to take long distance and close up looks at their world and discover the new things that can be seen through these special windows. This limiting perspective reveals beauty that is often hidden by the overwhelming visual stimuli in an environment. Focusing on one flower in a field of many helps children become aware of the little things that together make a whole.

What To Do

1. Choose an observation site. A familiar place might be particularly enjoyable.

2. Once destination is reached, provide each child with a viewer (paper rolls in a variety of lengths and sizes). CAUTION: Walking while looking through viewers can be dangerous.

3. Sit down and choose something to observe, such as a large tree. Sit close enough so that only a small portion of the tree can be seen through a tube.

4. Have a child focus on a spot and describe what is seen. Can others find it?

5. An adult should record observations on paper or tape for later comparison, discussion, and reconstruction of the object. Move closer or farther away and repeat the activity. What changes?

Want To Do More?

Find a spot and explore it close-up with a tube. Get closer with a magnifier. Get even closer with a microscope. Try long distance followed by binocular and telescope viewing. Take close-up and long distance photographs of the same object and match the part to the whole. Experiment with a variety of angles for viewing—everything from a bug's eye to a bird's eye view.

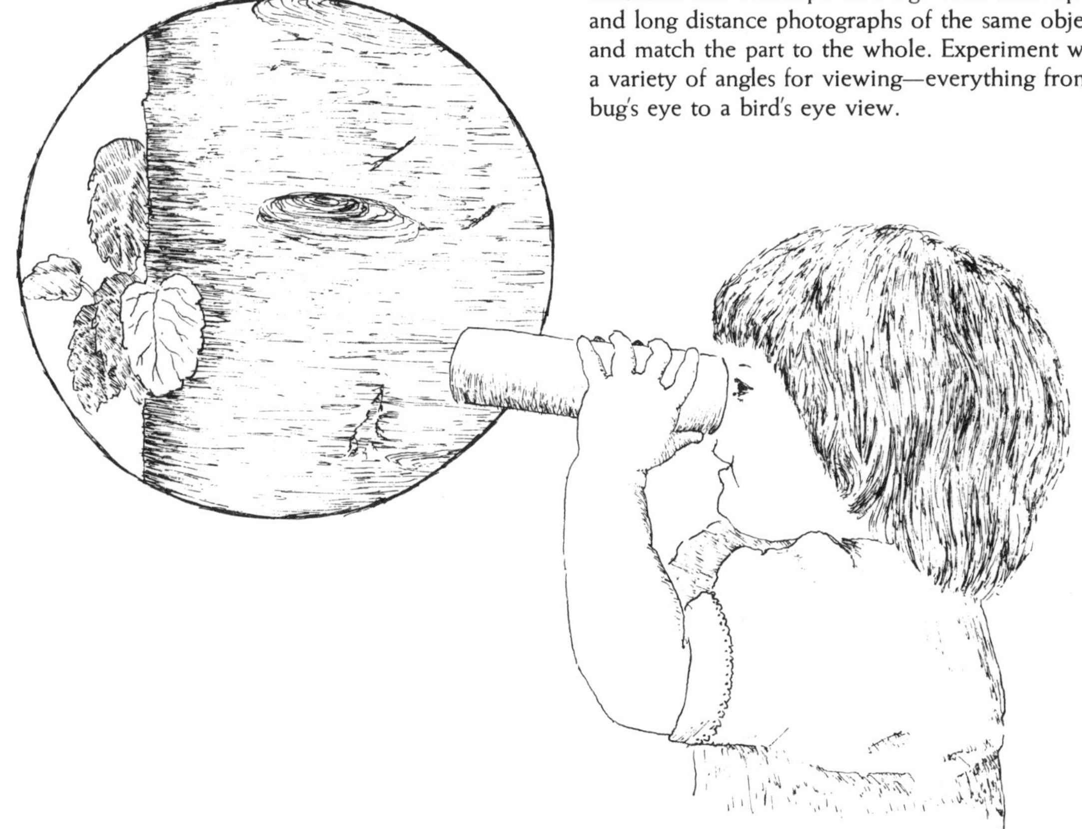

HUGGING A TREE

Things You Can Use

tree
blindfolds
note paper
pencil
drawing materials
tape recorder
tree identification book
tape measures

Words You Can Use

love
texture
senses
attribute
identification
rubbings
measure
meter

The purpose of this lesson is to develop a very personal feeling for the natural world. For many of us, nature exists as a vague whole rather than as a collection of unique parts, each important in its own right. By choosing a special tree to be a friend, learning its name, its likes and dislikes, its individual characteristics, a child can begin to feel more at home with nature. We all feel good when we enter a room full of people and see a familiar face. Children feel the same pleasure entering a woods and recognizing a friend. This personification can help children learn to love the woods.

What To Do

1. Choose your spot (a tree per child). If you have a large group of children, split the group as necessary. (Afterwards you could have the children share findings with persons who have selected the same tree.)

2. The children each go to their selected tree, give the tree a hug and introduce themselves to the tree. Each will then explore and observe their tree in every possible way, i.e. measure it, draw it, take rubbings, collect leaves, twigs, etc. Find the tree's name from an identification book or from the teacher. Record as much data as possible.

3. Tell the tree what you have found out about it and anything that the tree should know about you. Give the tree a "bye-hug."

4. Upon your return, each child can make a tree book. This book will contain all the personal data about their tree.

5. Return periodically to the trees. Each return should involve a hello and a goodbye hug and appropriate dialogue. During these visits, note changes in the tree and record them in your tree's book.

Want To Do More?

Follow a flower through the year. What happens to it in the winter? Does it return in spring? Try a shrub, a weed or a cactus. Plant a bulb and follow its growth. Visit a tree nursery. Find out what you can do to help your tree if it's sick or hurt. Have insects or pests attacked it? Find your tree's babies and help them to grow. Plant them somewhere else. Do you use trees for firewood? What can you do to replace the ones you burn? In what other ways do we use trees? Take a blindfolded friend on a tour of your tree. After removing the blindfold, can your friend find it? Find your tree's brothers and sisters (i.e. others of the species). Get to know your tree's friends. Are any of them endangered species?

A WOOD CHIP GARDEN

Things You Can Use

wooden discs
moss
sand
white glue
miscellaneous twigs
fungi
rocks
dried wood

Words You Can Use

terrarium
roots
appropriate plant names

Many people have had the experience of constructing a terrarium with greenhouse plants. Some have collected their own plants to replicate natural environments. An easy and inexpensive spin off from the terrarium is a wood chip garden. With a minimum of materials, children can take home part of an area visited to enjoy for a long time. We've even known a few to survive for years!

What To Do

1. The wooden discs are the base for forming your garden. Any wood slabs or bark pieces will do. Unique gardens are made from wooden discs cut from logs. For this you need to use a table or band saw. Cut the discs 1–2 cm thick, 5–10 cm across. Do this without the children's help.

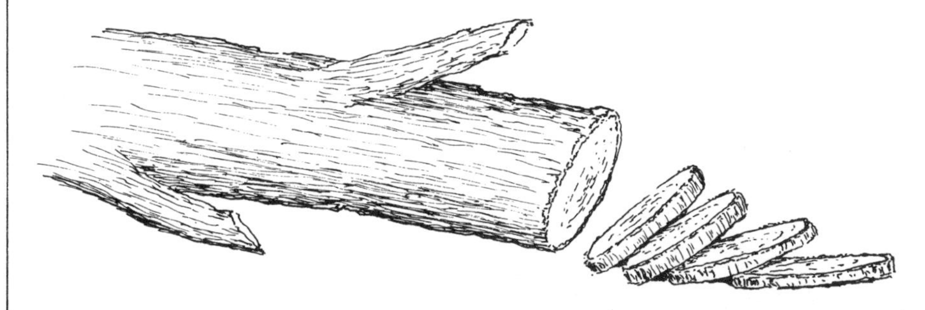

2. Collect moss or lichens from your area. Also collect sand (not beach sand, it's too salty and will kill the moss). River sand or sand from a construction site is best. Assorted rocks, twigs and other things may be collected for decorating the garden.

3. Spread a layer of glue on the wood disc.

4. Sprinkle a thick layer of sand on the glue. Shake off excess sand.

5. With the moss, twigs and rocks plan the landscape of your little world.

6. Once the plan is complete, glue the items, including the moss, in place with small amounts of glue. Allow it to dry undisturbed and out of direct sunlight.

7. After several days of drying it will need to be sprayed with water to moisten the moss. The garden will stay green and grow if you continue to spray it and keep it out of direct sunlight.

Want To Do More?

Build terrariums which represent other environments. Discuss the fact that moss can grow on a wooden disc because it has no root system, while plants with roots would die. Discuss the importance of roots.

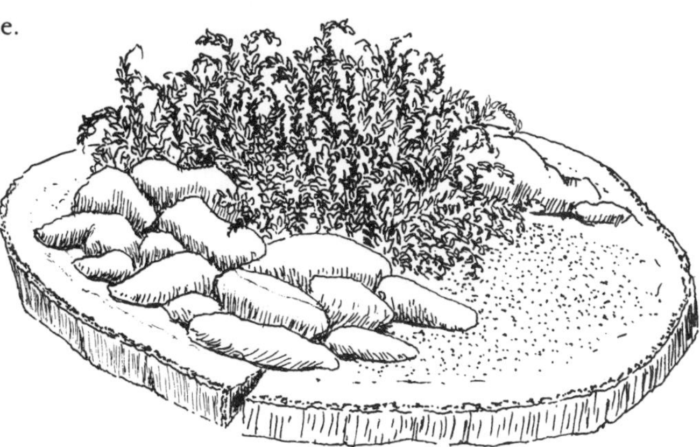

WHERE DO THINGS GO AT NIGHT?

Things You Can Use

flashlight
cardboard
talcum powder or corn starch
insect repellant
Sit-Upon (see pg. 100)
tape recorder
prepared tape of night sounds (Most public libaries have recordings of animal sounds.)

Words You Can Use

nocturnal
phases of the moon
full moon
quarter moon
constellations
plant and animal names as appropriate

There is a special thrill for young children in going out into the night. Even the very familiar world becomes a different, sometimes scary, place. It's like having a special secret to be outside exploring when most other children are inside or asleep. Although many children are a little afraid, the spirit of adventure usually wins out and they are eager to explore as long as there is a trusted adult along. Children are not often quiet enough to see an abundance of wildlife; however, you might see cats prowling, hear night birds, or find some flowers which close at night. A night walk also provides the opportunity for children to compare night life in the wild with their own evening routine.

What To Do

1. Choose a trail. The leader needs to prewalk the trail to choose a good spot for sitting and listening, a place for walking without flashlights, and possible observation sites.

2. Each child should be prepared for the hike with a flashlight and proper dress and shoes. If it is a damp evening, a sit-upon can be used (2 people per sit-upon).

3. Have children walk in pairs (holding hands is a warm human way of overcoming fears and building trust).

4. Begin the walk. Stop and listen occasionally. Try to have a quiet walk. Many children who fear the night will compensate by making much noise. As the walk progresses, this fear lessens.

5. Find a place to stop, listen, and rest. Play the recording of night sounds. Have the children listen for actual sounds, similar or different. Add these new sounds to your tape, if possible.

6. As you return, look for animals in the night. Do this by carefully watching the beam of your flashlight as it passes through the woods, grass or shrubs. Animals' eyes reflect the light. Watch for this reflection and then pinpoint the creature. This works for spiders, insects, birds and other nocturnal creatures.

7. With home just around the corner and the path now familiar, it may be the right time to turn off all the lights and walk home as a night creature would. This can be rewarding as the eyes become accustomed to the lack of light and more things become visible. It is interesting to note that eyes at night see no color.

Want To Do More?

Rewalk the trail with no lights at full moon or at no moon. Place a piece of cardboard on the trail or at a place animals might use. Cover the cardboard with a light dusting of talcum powder. Wild night animals will leave their tracks. Place food out at dusk near the powder. Identify animals from their tracks. If the stars are out, find clear places and observe.

Talk about phases of the moon. Explore the brighter stars and planets. Locate them on subsequent nights.

A campfire is a pleasant end to a long walk, and it may serve as a focus for your discussion of the night's activity.

TALKING ABOUT MONSTERS

Things You Can Use

tape recorder
camera
collecting bags
gloves
paper and pencil
drawing equipment
bird and animal recordings from the woods (or your public library)

Words You Can Use

poison
senses
explore
environment
reaction
allergic
imagination
reptiles

Choose words specific to your location that might cause fear or doubt about a site, i.e.
- full moon
- coyote sounds
- night bird sounds
- marsh
- swamp
- fog
- mist
- shark

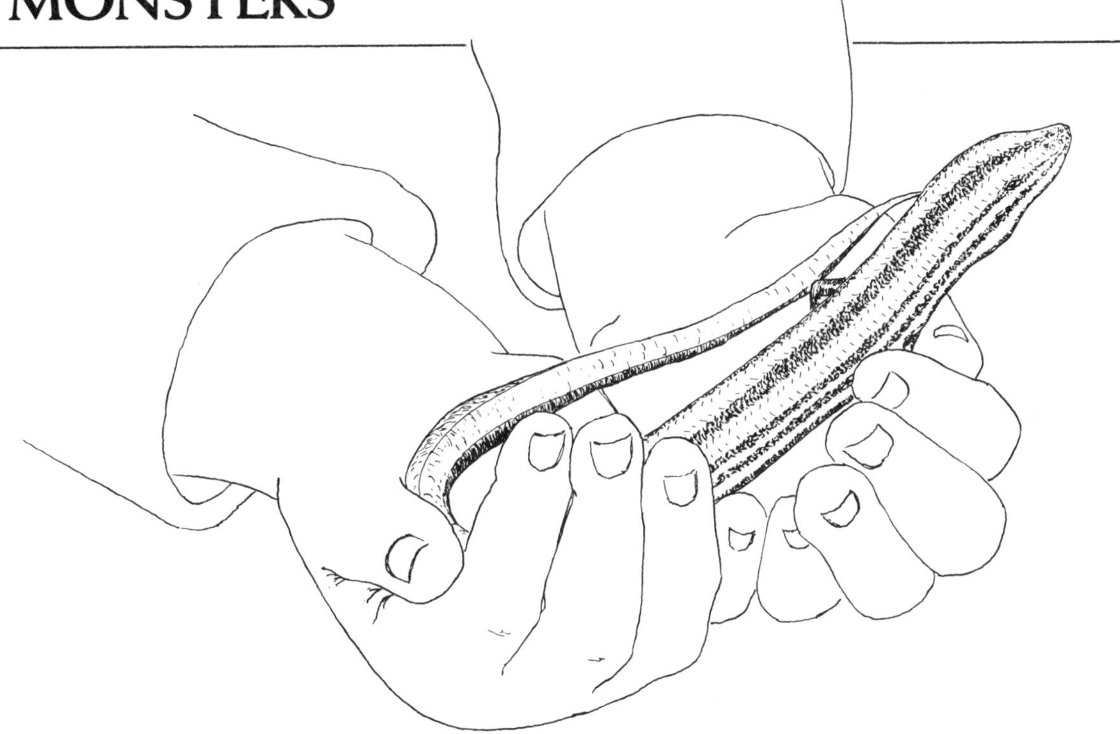

Monsters in the wild aren't always big and hairy. They can be mud and slime, poison ivy, eerie sounds, dark places, or snakes and spiders. Some of them we should be wary of, while others are monsters simply because of our own conditioning. Caution is appropriate in some circumstances, but caution and fear and prejudice ought to be differentiated. In this lesson, we'd like to separate the real from the not-so-real monsters and learn a bit about both.

Adults need to be aware of any fears or misunderstandings which they themselves have so that, if possible, they can avoid passing these on to the children. Coming to grips with anxious feelings will allow the children to explore freely and to develop their own attitudes about the outdoors rationally.

What To Do

1. Discuss with the children things and situations which have caused fear or anxiety. What do they consider to be repulsive ("yucky")? Discuss the "whys". Share your own experiences. Discuss situations in which caution is wise, such as when approaching certain animals or entering unfamiliar areas. Encourage children to draw or write about their monsters, whatever they may be.

2. As a result of these discussions, the adult should choose a site for beginning exploration. Potentially uncomfortable situations should be discussed ahead of time and appropriate precautions taken. These might include special clothing or hiking guidelines. Records can be used to familiarize children with sounds of the outdoors.

3. Go to the site you have selected and explore it as thoroughly as possible. What do the children find which they feel is harmful? When something is found, share it with everyone. This is the point at which reality and imagination need to be differentiated. For example, everyone knows that some spiders are dangerous. A spider in its web is not. It can be approached with a magnifier. It can temporarily take up residence in a jar for further

viewing. This can also be the time to separate surprise from fear. No one likes to be startled by something unexpected, such as the feeling of a spider's web on the face. That is different, however, from being afraid of spiders in general. Again, it is true that some spiders are dangerous, so all should be approached with caution.

4. Following this experience, talk about the difference knowledge can make in your feelings about exploring the environment. Is there a difference in your feelings as a result of this exploration? Record your findings and observations by whatever means are most appropriate (i.e. camera, tape recorder, notes, drawings).

5. Start planning for your next monster encounter.

Want To Do More?

The approach outlined about can be used in dealing with a variety of fears and anxieties, such as, fear of the dark, thunder, and lightning. If possible, photograph your monsters. How do you feel about the developed pictures? Are they more or less scary than the real thing? Make a tape of scary sounds. Make up a story about your monsters. Use the tape for sound effects. Illustrate it.

Yucky things, while not actually feared, can be just as big an obstacle to exploration as scary things. Children can learn to cope with them when necessary, especially if prepared, in order to discover a previously untouched area such as a swamp.

WHAT MAKES A PERFECT DAY?

Things You Can Use

drawing and writing materials

Words You Can Use

agree
common
likes
dislikes
perfect
weather
season

What's your favorite season? Your favorite kind of weather? What makes a perfect day for you? In this activity children focus on what they most enjoy doing outside and how it is affected by environmental conditions. It gives children a chance to say what is really special for them and to act upon it. It also makes them more aware of the feelings and ideas of others. This is an activity which gets everyone into the act on an equal footing. Besides that, it's fun!

What To Do

1. Ask everyone to draw a picture or dictate a story about what they feel makes a perfect day.

2. Discuss the pictures and stories. Make a list of the things that make perfect days.

3. Plan and take several field trips that will include those things discussed that make the perfect day (i.e. a hot day for your swimmer, the first snow for your sled rider, a birthday day with no rain). There should be many common items or environments, and there will be a variety of tailored field trips that will be enjoyable for all participants.

Want To Do More?

Do this activity for each season for the year. This will provide an interesting variety of trips. Discuss the reasons for the choices. What influenced the choice—past experience, something your parents have talked about, something you heard in a story?

A PIECE OF OUR WORLD

Things You Can Use

knife
spade
garden trowel
small plastic sandwich bags
magnifying glass
microscope (a children's microscope is available through Childcraft Education Corp., 20 Kilmer Road, Edison, New Jersey, 08817)
poster board
scotch tape
scales

Words You Can Use

soil
earth
samples
same
different
abuse
care
love
heavy
light

Soil is one of the major supports of life, yet both children and adults tend to think of dirt as just "dirt." In this activity we ask children to collect and examine "dirt" from a variety of settings. In this way, they observe the effect human activities have on the soil. Going out into the field, collecting samples, and bringing them back for analysis has a lot of appeal, and it gives children a model for further formal and informal science experiences.

What To Do

1. Discuss a variety of environments, stressing the commonality that they all have (soil).

2. Take a walk that will expose the children to a well groomed lawn, the woods, a hard-packed playground, a roadside shoulder, a plowed field, a window box, a garden, or other available settings.

3. Use the garden trowel, spoon or fork to gather soil samples from each site. Place samples in sandwich baggies, seal with tape and mark with the name of the site.

4. Take samples back home or to school.

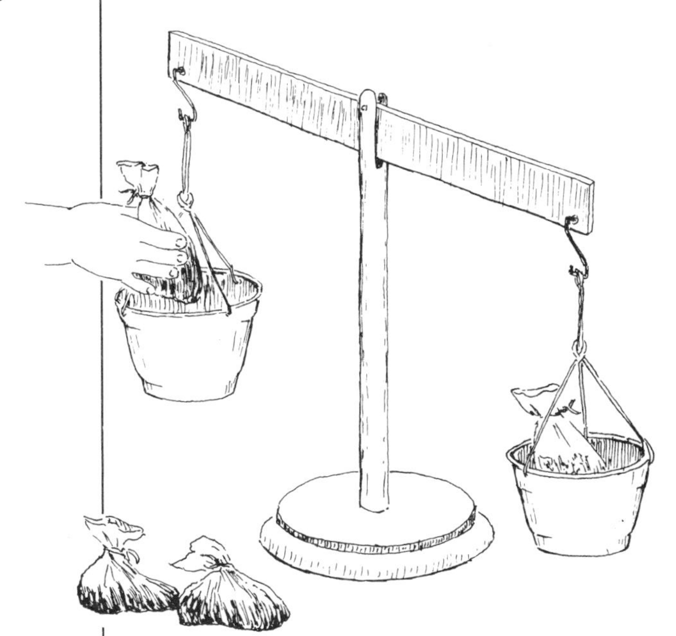

5. Examine samples from each site. Sort soil by categories such as color, rocks or no rocks, worms or no worms, or wet or dry. Compare equal quantities of soil using a balance scale (i.e. a soup can full of each type). With children who count—how many inch cubes does each can full weigh? Record your information following the suggestions in Chapter I.

6. Discuss what makes the different samples look different. What have we done to it to make it the way it is? (i.e. trampled a foot path, plowed and fertilized a garden)

Want To Do More?

Examine additional sites. Write stories with the children about how you feel you could take better care of your world. Draw or paint pictures that depict the sites visited and how they were different from each other.

Take a stake. How many hammer strokes does it take to pound it into the ground? Try it in another area.

For those in a classroom setting, ask each child to bring a soil sample from their home for study and comparison.

Discuss or read stories about why soil is so important to the farmer, the home builder, the highway engineer, the gardener, the nursery man. What would the child's world be like if all soil was like sand from the beach?

EVERYTHING IS CONNECTED

Things You Can Use

paper and pencil
cards
string
identification books

Words You Can Use

interaction
web of life
ecologists
predator
prey
consumer
producer
living
non-living
decomposer
food chains
habitat

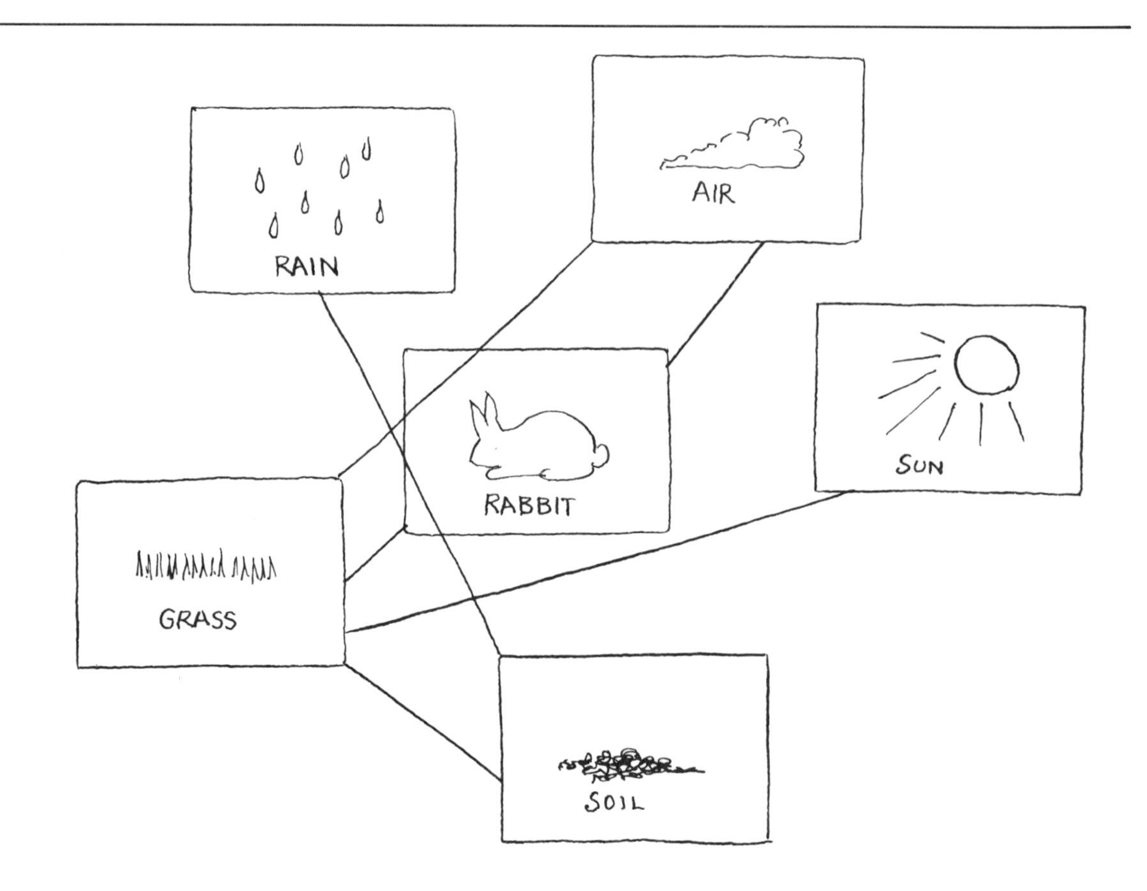

Each object in an environment interacts with every other object in some way. To this idea every ecologist will attest with certainty. Children, however, may not be able to draw relationships between the various components of an environment. Identifying these interactions in a familiar environment is one way of exposing children to this concept.

What To Do

1. Choose an environment (i.e. school yard, beach). Define the boundaries. It's wise to begin small.

2. Develop a list of all objects in that environment —everything from dirt to kids. Make a card for each, with the name and a picture, if needed. The older the kids, the more complex the list will be. An example from a backyard might be: soil, rock, worms, grasshoppers, dandelions, dead leaves, trees, squirrels, birds, people, dogs, a can, a piece of paper, rain, sun, air.

3. Form your area's web of life. This is done by giving each card to one person to hold, and then running a string from each card to every other one with which it interacts in the environment, thereby forming a complex web. Alternatively, you could lay the cards on the floor or put them on a bulletin board. What happens when you remove one card? How many ways can you arrange your web?

4. Discuss your findings. What are the most influential factors? Least influential? Where do humans fit in?

Want To Do More?

Explore a mini-environment such as under a rock, a rotting log, a small pool of stagnant water, a bird nest, a human home. Develop a web of your own with magazine pictures. Try to make it complete. Choose an animal. What does it need to survive? Classify the cards by whether they're producers (green plants), consumers (plant eaters), predators (animal eaters), prey (it's eaten), or decomposers (rots things—mushrooms).

Draw a picture chart of a food chain for an animal or a plant. What supplies it with food? How does it provide food?

Picture the mini-environment as a "home" (habitat) and compare it with your own home environment. What do we need to survive? Make a picture chart to illustrate.

HOW MANY WAYS TO GRANDMA'S HOUSE?

Things You Can Use

paper
pencil
crayons
a watch (or stop watch)
map
camera
tape recorder
compass
collecting bags
trundle wheel
a destination

Words You Can Use

map
time
north
south
east
west
meter
environment
direction
compass

There are usually a number of ways to travel to a given destination. Even young children are aware of this. Exploring alternative routes to a destination provides an opportunity for the acquisition of new skills as well as an appreciation for the variety that exists within a limited environment. Emphasizing the aesthetic aspects of everyday routes may help children understand, in a more general sense, that all environments have something to offer.

What To Do

1. Choose your destination.

2. Choose one route for going and one for your return. Walk them.

3. Compare the two routes. Which way did you like best? Why? What other ways can you go?

Want To Do More?

Time the routes. Measure the routes with a trundle wheel. Map the routes. Use a camera, tape recorder, and/or drawings to record information about your walk. Make collections along both routes and compare them. Rank routes according to a selected priority—scenery, speed, ease of travel, safety, and others. Which route is easiest to travel? Make up directions so another person can follow your route. Why do roads and paths go where they go? What physical features are involved? How can you adapt these activities to a long trip? What special sites did you discover? Take a friend on this walk and share your discoveries.

TURNING ON YOUR SENSES

Observation Experiences

TOUCH ME, FEEL ME, KNOW ME, OR WAKE UP YOUR FINGERS

Things You Can Use

collected items from the outdoors, i.e. rocks, leaves, soil, twigs, objects similar in size but different to touch
paper bags

Words You Can Use

texture
touch
feel
similar

How frequently we adults use the words "Do not touch" even though we know children must handle things to learn about them. In this activity, children are urged to explore with their hands. "Do touch" becomes important. Children handle various items from their surroundings and identify, sort, group, and classify them. In addition, their vocabulary grows as they seek the words needed to describe what they feel.

What To Do

1. Show the child two items very different in size, shape and texture. Put items in the bag. Let the child reach in and touch one object. As the object is touched and felt, the child identifies it.

2. Add more items, 4 or 5. Repeat the above procedures.

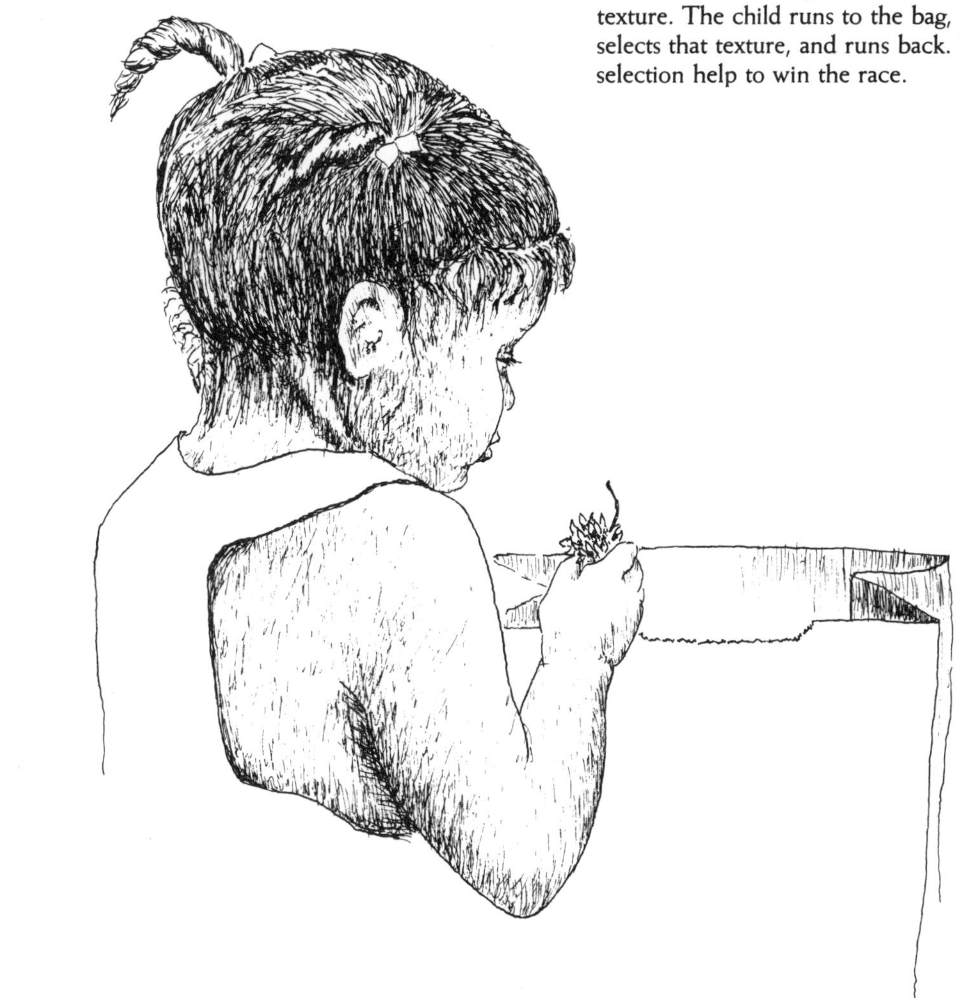

Want To Do More?

Have the children look for things to use in this touching game. They can find more things to use from outdoors or from home.

Let the children exchange bags and try to identify each other's choices. Have the children describe what is in their bags for others to guess.

Run a texture relay. The teacher names a texture. The child runs to the bag, reaches in, selects that texture, and runs back. Speed and selection help to win the race.

OPEN HOUSE

Things You Can Use

pictures of animal homes

Words You Can Use

habitat
nest
cocoon
den
web

Children enjoy watching the animals they encounter in their surroundings, but very few have a real idea of where those animals live. They may talk about squirrels "Going home to their mamas" but still have no clear mental image of that home. Through searching with an adult for squirrel nests, spider webs, or ant hills, and discussing what they find, children can broaden their concept of "home" and learn some of the reasons animals choose the homes they do. The experience also exposes them to the value of close and careful observation.

What To Do

1. Prior to the trip the adult must locate animal homes in the area to be used.

2. Begin with a discussion about people's homes. Move on to the idea that animals have homes, too. Some are permanent such as a fox den. Others, such as birds' nests, are for raising young and are abandoned quite soon. A spider web, an insect gall, and a rotten log are all homes for some woodland creature.

3. Take the children out to see the homes. Take time to observe the animals and talk about how we can protect or harm the homes. You may also want to talk about how some animals' homes may need to be removed (i.e. bagworms), and for what reasons. Is it harmful to you or to your plants? Note: All animal homes in the wild should be left undisturbed.

Want To Do More?

Help each child find a home in the designated area; he or she should learn as much as possible about the animal living there. Mark the spot with the marker given. When all the children are finished, begin an open house tour. As each marker is approached, the child who placed the flag is to step from the group and conduct an open house explanation of the animal's home.

Ideas for Older Kids:

Put numbers at each site and write up or tape an open house dialogue. From the local real estate person collect details and descriptions of a real house's open house. Try to model your open house dialogue after the real estate ones. Invite in a real estate agent. Draw a map of your homes; note location and kind of dwelling. Note seasonal changes. Discuss how animals select a home and how we select one.

Explain how loss of habitat causes certain creatures to be killed because they have no place to live. An example is the disappearance of woodpeckers if no standing dead trees are left.

1, 2, 3 HUSH

Things You Can Use

proper dress
sit-upon
tape recorder

Words You Can Use

auditory
quiet

Kids often invade the wilds with the same vigor as they do the playground. There is running and jumping accompanied by shouts and exclamations: "Look what I found!" "What's this?" Let me see, too!" The adult participants, while less vigorous in running and jumping, often out-do the children with continual comments urging them to look at things. But what about the slight trickle of a rivulet, the gentle chirp of a song sparrow, or the wind in the trees? While we want to encourage visual observations, there is a world full of sounds which shouldn't be missed. The purpose of this activity is to emphasize the value of listening to gentle sounds as a means of getting to know our world.

What To Do

1. This activity requires a walk to a site that is fairly isolated from human activities.

2. The walk should be a normal exploration activity with the children involved in visual exploration.

3. As you approach the isolated site, the group should be stopped and asked not to talk and to walk quietly.

4. Using this orderly quiet walk, move to the isolated location and find a spot where the group can sit and listen. The key here is a comfortable position for each participant: lying flat on the ground or sitting and leaning against a tree are usually best. On the count of 3 every one stops talking and stays very still. For two to three minutes listen to the sounds of the place. When the quiet time is finished discuss what was heard and the children's feeling about the experience.

Want To Do More?

Compare the sounds to music. What could be used to create similar sounds? What sounds did each person like? Compare natural to man-made sounds. How did each make you feel? Try the listening activity in different areas. Make lists of the sounds in different places and compare them Record various sounds. How do you feel when man-made sounds invade the quiet natural environment? Make one minute tapes in various settings. Can you identify them later? Do sounds matter to wild animals? How? Imagine that you are a rabbit. What would the sound tell you? What sounds are important to a hunting animal (fox)—to a hunted animal (rabbit)?

MAKE A RAINBOW

Things You Can Use

garden hose
water mist

Words You Can Use

names of colors
rainbow
water
spray
mist
sun
sprinkle

A rainbow is a beautiful arch of colors that appears in the sky when the sun shines after a shower of rain. Seven colors appear in every rainbow: red, orange, yellow, green, blue, indigo and violet. Most of the time we can only see four or five colors clearly, since they blend into each other. You don't have to wait until it rains to see a rainbow—you can make your own. This activity gives the children an opportunity to create, observe, and enjoy the fun of their very own rainbow.

What To Do

1. On a warm sunny day, let the kids put on their swimming suits.

2. Turn on the garden hose. The kids will love running in and out of the water.

3. After the initial enjoyment of getting wet and cooling down, adjust the hoze nozzle to a fine spray setting.

4. Arch the spray of water high into the air.

5. Let the children observe the rainbow that is created as the sun's rays strike each water droplet. The millions of small water prisms form a giant rainbow.

6. Let each child take a turn creating a rainbow.

7. Discuss a rainbow, what it looks like, when it appears, and the many colors that are in it. It would be helpful to use a picture of a real rainbow. Compare the colors seen in the rainbow to those that were in the picture. Are all the colors the same? Are some missing? Isn't it great to make a rainbow?

Want To Do More?

Make a picture using all of the colors of the rainbow. Use a prism to show the colors of a rainbow. Ask the children to tell you about their feelings when they see a rainbow. Does a rainbow make you sad? Happy? Why?

Talk about the folklore of the rainbow. Find a rainbow story to read to the children. Can you find a pot of gold at the end of your rainbow?

Change the nozzle to adjust the water volume and the size of the drops. Can you still make a rainbow?

LET YOUR NOSE DO THE WALKING

Things You Can Use

your nose

Words You Can Use

smell
odor

There are many different kinds of smells both in your home and out of doors. This activity gives the children an opportunity to turn on their noses to find out just how many different odors they can detect both in the home and outside. They will develop an awareness of the variety of smells in their world.

What To Do

1. Take the children on a slow walk through the house or classroom. Ask them to use their noses and to tell you each time they smell something.
2. Try to identify each smell with the children.
3. After the walk is completed, discuss all of the odors.
4. Repeat the above procedure only this time go outside.
5. Discuss and compare the odors inside and outside.

Want To Do More?

Sensitivity to odors can become fatiguing. In other words, our noses get tired, too. This might happen on the walk. Talk about it. Some smells last longer than others. Collect a smell in your hand, i.e. dandelion, sassafras leaf. How long does the smell stay in your hand?

THE TEXTURE COLLECTOR

Things You Can Use

collecting bags
board for gluing specimens of texture
blindfold

Words You Can Use

texture
touch
soft
smooth
rough
hard

The young child who has felt the soft pile of a forest's mossy bed, the cold hardness of a granite cliff, or the rough jagged heads on a barnacled board, knows that texture is an integral part of an environment. Experiences with texture are often overlooked. In this activity, children have the opportunity to follow their natural inclination to collect things as they walk. This time, however, the focus is on collecting textures.

What To Do

1. Prepare a piece of poster board by dividing it into squares (20 cm × 20 cm).

2. A texture from nature will be placed in each square.

3. The word texture should be defined for the children, either on the trail or after returning.

4. Go on a collecting walk. The object of the walk is to gather as many textures as possible. Remember that you must return with the textures in a form which can be placed on the texture board.

5. Glue the collection onto the texture board. The items may be placed at random, or grouped into categories such as rough, bumpy, or smooth.

6. Blindfold the children and ask them to describe what they feel.

Want To Do More?

Take a texture walk along the collecting route and identify the sources for your texture board. Find textures that are similar to those on the texture board. Make a set of texture rubbings using chalk or crayons. These can be matched to the texture board. Make a texture lotto game from your rubbings. Try to develop connections between textures and emotions. What is your reaction to touching a rough object? A slimy one?

Glue texture pieces on 3 × 5 cards in pairs. Match the pair of cards while blindfolded or while holding them behind your back.

Make a list of texture words.

SOUNDS THAT MATCH

Things You Can Use

33 mm film containers or margarine tubs
pairs of matching objects to fit into the containers, i.e. water, sticks, rocks, bark, seeds, acorns, pine cones, nuts, pine needles

Words You Can Use

pair
soft
loud
alike
match
together
describe
container
similar

The world is full of sounds. We can all identify the common ones in our homes—the telephone, water running, a vacuum cleaner. It takes careful listening to identify nature's sounds, which are often quiet and subtle. This is an activity which will develop and sharpen auditory perceptual skills as the children match sounds which are alike. By becoming more precise listeners, children will be able to tune into a world of sounds which may previously have gone unnoticed.

What To Do

1. The child shakes a container and listens to its sound.

2. The child shakes other containers until one is found that sounds exactly like the first.

3. The child puts the two containers that sound alike, or match, together.

4. The child follows the same procedure for the rest of the containers.

Want To Do More?

Discuss pitch of sounds (high-low). Discuss what the child thinks is making the sound. Discuss where the sounds may have been heard before. Let children open the containers to check their hunches. Let children make their own sound containers. Put the same objects in different containers. What changes do you notice? What other sounds of nature can you capture?

Pass out one container to each child in the group. Then give them the task of finding their partners by searching out the matching sound. This is especially effective if there is no talking.

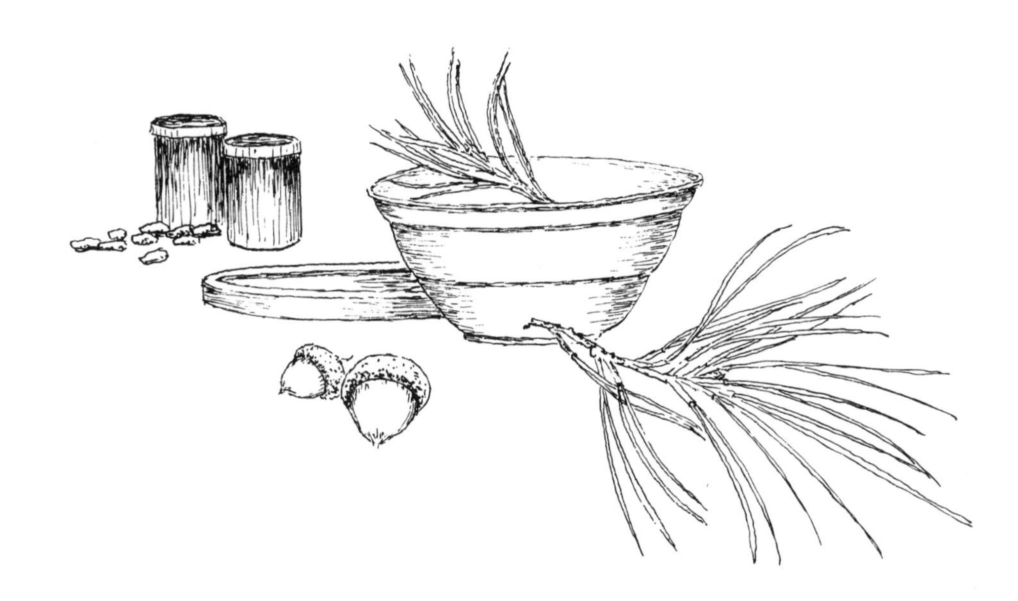

FOLLOW THAT CRITTER

Things You Can Use

your eyes

Words You Can Use

appropriate animal names
hop
crawl
run
wriggle
fly
swim

It's fun to pretend to be a critter from the woods, to eat as it eats, and to walk as it walks. But, in order to imitate an animal's actions, we must first observe it very carefully. We should follow its movements and watch its behavior. Some will move very small distances (ants, pillbugs) while others (squirrels, birds) will cover much ground. To know how they move one must watch. This activity encourages children to do just that, watch a "critter."

What To Do

1. Help each child find an animal to follow. Possibilities include ants, caterpillars, beetles, roly polys, birds staying near a feeder (not those just flying by), fish in a clear water source, worms, toads, squirrels, lizards.

2. Observe your animal. If possible, follow it for a while. Watch it carefully to see how it moves. Does it stop to eat? Where is it going? Are there others of its kind?

3. Come back together and talk about the observations. Let each child move like his or her animal. Can the rest of the group do it, too? Where is its home? Is it part of a "family" (social group) or alone?

Want To Do More?

Very gently disturb the creature. What does it do? What happens if you move suddenly? Check the spot at various times of the day. Do you always find the same kinds of creatures? Do they always do the same things?

Pretend you are that animal. What do you think it is dreaming about and planning to do next? What things does it do that you and your family do (i.e. eat, sleep, play)?

LEAVES DON'T ALL FALL THE SAME WAY

Things You Can Use

a wooded area with a variety of tree species

Words You Can Use

tree names such as oak, maple or elm.

Note: because this is a highly verbal activity, teachers may need to supply descriptive terms for the actions of the leaves

Big leaves, little leaves, yellow, orange, and brown leaves become as jumbled in our minds as they do in the autumn leaf piles. Millions of trees end the growing season by dropping their leaves to the ground in preparation for the winter's rest. Exploring the great variety of ways that they fall is a novel means of becoming familiar with leaves, and can lead to numerous other activities such as classification and body movement.

What To Do

1. Find a location under a tree where you can lie down. Choose a calm day in the fall when the leaves are falling.

2. Watch as the leaves fall naturally. A few will be falling most of the time.

3. Describe the ways that they fall. Record the descriptions.

4. If possible, shake the tree so that many leaves fall. What new fall patterns do you notice? Describe these and record them.

5. Move to an open place such as a lawn. Have each child describe and, through movement, act out a falling leaf.

6. Then, let the entire group pretend to be a forest of falling leaves.

Want To Do More?

Go to different kinds of trees; see if the leaves fall in different ways. Using a rake to gather the fallen leaves, cover each child in turn, or several children, depending on the size of the leaf pile. Describe how it feels to be a fallen leaf, or to be hidden in the pile. How do fallen leaves help the forest? Discuss why leaves fall. Point out that some trees drop their leaves in fall and others do not. Do all leaves fall sooner or later?

Have each child choose a leaf to drop from a high place. What happens to leaves after they fall? i.e. leaf mulch as winter protector for insects and animals, soil component.

CURVES AND STRAIGHTS— THE SHAPES NATURE MAKES

Things You Can Use

shapes drawn on cardboard–shapes should include drawings of a circle, a triangle, a rectangle, and a square
glue or paste
poster board

Words You Can Use

circle
rectangle
square
triangle
shape

It is the purpose of this activity to allow children to discover that circles, triangles, and squares are not forms created by teachers or parents. The real creator of these shapes is the natural world. Searching for these shapes in nature reinforces the identification of the basic forms and promotes visual perception skills.

What To Do

1. Give out shape cards.

2. This is a hunt. Find a shape in nature that closely resembles the shape on the card. Collect the objects (nuts—round, redbud leaves—heartshape, bark—often rectangular).

3. When all the shapes have been found, go to a spot where you have laid out glue and poster board. A section of poster board should be used for each shape.

4. Each child pastes his or her shape on the appropriate board for a shapes collage. You can add to this as new items are found.

Want To Do More?

Find objects that have more than one shape. Have a shapes bingo game. As the shapes are found add them to the bingo card. Find objects to match solid figures such as cylinders, boxes, cubes, balls.

Place an object in front of a light to determine the shape of its shadow. Try to predict the shape it will cast. Move the object around. The shadow may change shape.

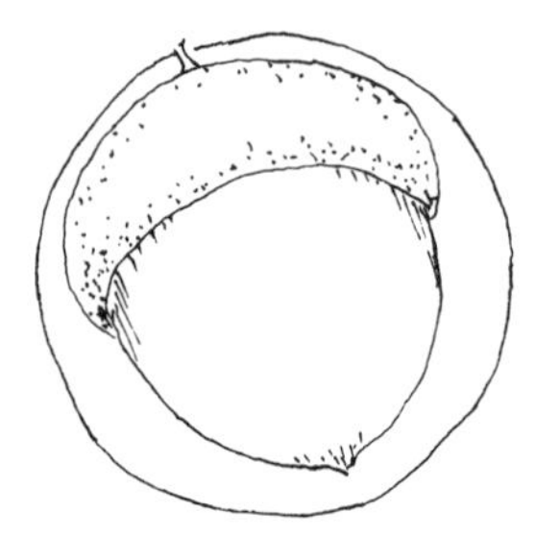

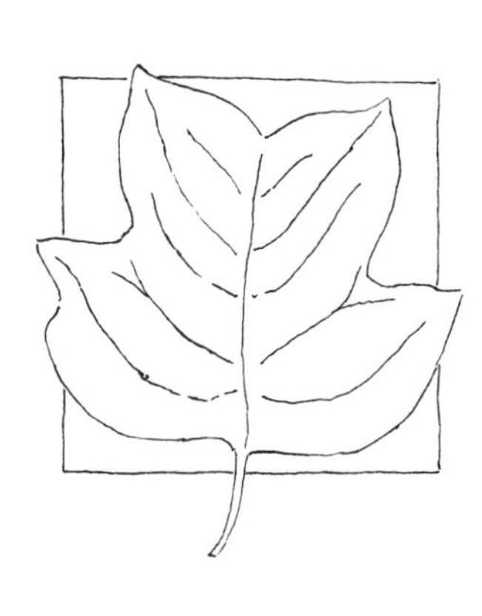

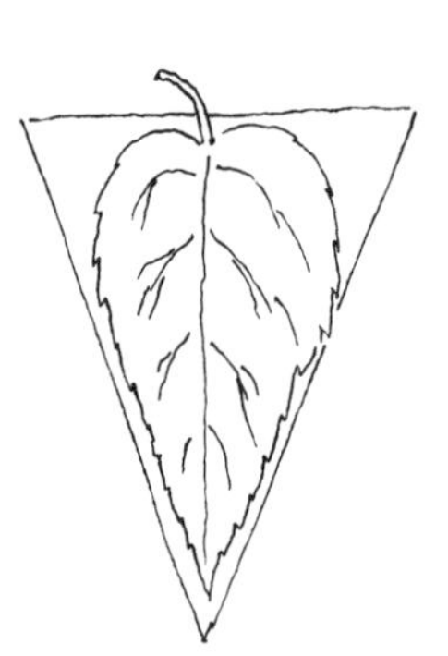

TAKE A BIRD TO LUNCH

Things You Can Use

an empty plastic milk or bleach bottle
or a small wooden board with molding of some type nailed around the edges to prevent the food from blowing away
a water dish
grains
dry cereals
all types of seeds
cracker crumbs
etc.

Words You Can Use

feeding station
bird feeder
identify
appropriate bird names

Our friends the birds need to eat just as we do. Tending a bird feeder helps the children develop an interest in and a love for birds. It also provides an opportunity to identify different species of birds, their style of feeding and what they like to eat.

What To Do

1. Build a bird feeder—see Hodge Podge for construction of feeders.
2. Attach the feeder to the outside ledge of the window, or on a visible tree branch. Place different kinds of food on the feeder. Put a water dish out as well.
3. Caution the children not to make sudden movements as they watch the birds feed.
4. The adult should identify the birds for the children using pictures and manuals.
5. Observe the kinds of foods different birds eat.
6. Observe the total number of birds that come to the feeder.
7. Observe how long different birds stay at feeder.
8. Observe the ground feeding birds.

Want To Do More?

Build more kinds of feeders.

BACKYARD DATA COLLECTING

Counting, Classifying, and Recording Experiences

HIDE AND SEEK FOR CRITTERS AND KIDS

Things You Can Use

area to explore
writing materials

Words You Can Use

observe
examine
explore
search
record
predator

Animals need to hide in order to protect themselves from predators and the weather. Just as children like to play hide-and-seek in areas with lots of hiding places, animals live where there are many safe places for them. In this activity, children explore an area for their own hiding places. Then they pretend to be an animal and look for hiding places. It soon becomes obvious why even though deer like grass, they don't live in vacant lots!

What To Do

1. Walk around the yard of your house or the playground at school with the children.
2. Observe, count and record the various places that a child might hide when playing the game hide-and-seek or kick-the-can.
3. Choose an animal. Observe, count and record the many places that that animal might hide.
4. Discuss the idea that hide-and-seek is fun for children.
5. Discuss the need for animals to have hiding places. Point out the implications if the children do not see them.
6. Compare the number of human hiding places and the number of animal hiding places.

Want To Do More?

Can you find any evidence in a hiding place to show that an animal has been there? What evidence do you leave behind where you hide?

Name some of your favorite animals. Discuss places that they hide and why. How does the lack of hiding places affect an animal's choice of habitat? Are the animal's shape, color and texture important?

What places can you hide at night that would not hide you during the day? Does a creature have to be out of sight to hide? (The answer is no. A squirrel may be safely hidden in a tree, though still in view.) How do baby animals hide?

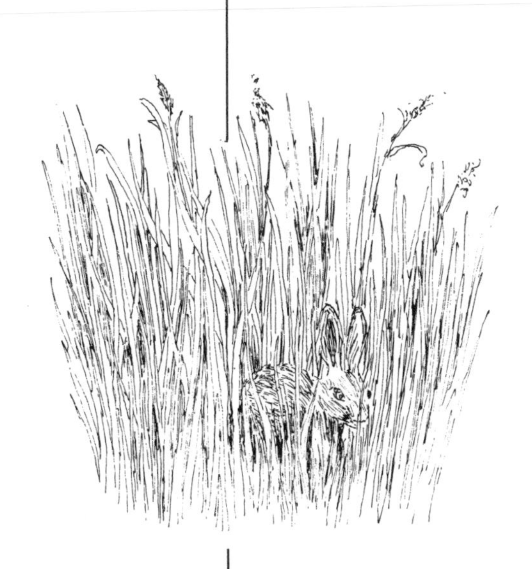

GROW A SOCK

Things You Can Use

long socks with fuzzy outer surfaces to which seeds will stick (i.e. adult knee socks)

Words You Can Use

dispersal
seed
germinate
plant names as appropriate

Collecting seeds and nuts is a natural activity in the fall. However, a collector often overlooks many seeds because they are small or hard to recognize. An entertaining way to collect some hard to find seeds is to take a sock walk. Previously unnoticed seeds will be easily collected and as a bonus, one method of seed dispersal will become very obvious.

What To Do

1. Dress each child in a thigh high pair of socks.

2. Go for a walk through a densely vegetated area. An empty lot overgrown with weeds would be excellent.

3. Return to home or class and look at the socks! Then take them off.

4. Wet the entire sock, and place it in a cake pan placed on a slant. (see illustration) Fill the lower portion of the pan with water so that the sock remains wet.

5. Put the pan in a warm place and watch the seeds sprout.

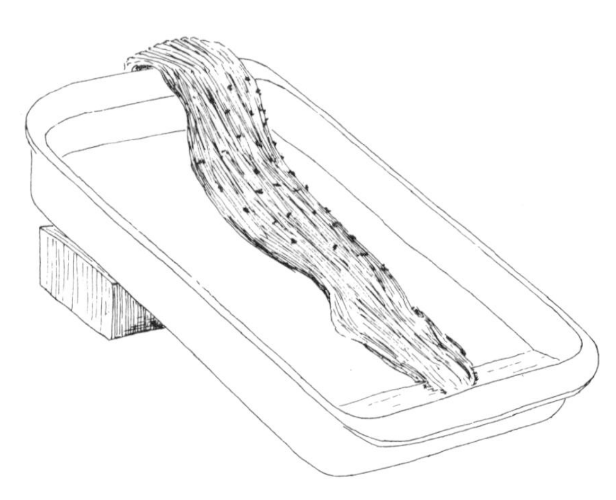

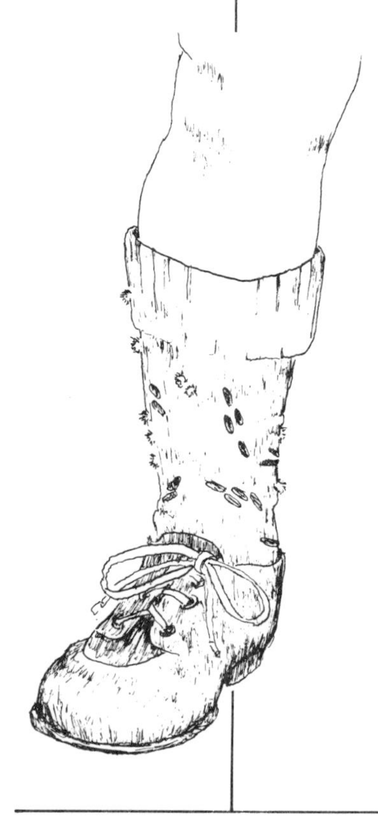

Want To Do More?

Pull the seeds off the socks. Sort and place them into cups by species. Allow them to dry. Divide each cup of seeds in half. Place one half in a freezer for 2 weeks. This is to simulate winter. Some plants won't grow without freezing. Next, plant seeds from both halves in "Seedbed in a Bag" (See "Hodge Podge").

Take sock walks at different seasons. Which seeds are harder to remove? Do some hurt you? Can animals help seeds find new places to grow? Yes! Glue samples of seeds on cards to develop a seed collection. Repot sprouts and grow them to full size.

What other ways does nature have of spreading seeds around (e.g. winged seeds—by wind, berry seeds—by birds)?

Plants with fur carried seeds need animals to make sure they are widely spread. Do you think the plants do something to help animals in return (provide food, shelter)?

FEED THE CRITTERS

Things You Can Use

water dish
grains
dry cereals
all types of seeds
cracker crumbs
raw vegetables, etc.
recording materials
  i.e. pencil, paper, tape recorder

Words You Can Use

observe
record
identify
time

Birds, squirrels, and other animals live in every backyard and playground, coming and going in their own daily rhythm. The provision of a food source allows these critters to be observed daily and recognized as a consistent part of the environment. In addition, it provides the opportunity for their patterns of behavior to be discovered and perhaps related to the behavior of the children.

What To Do

1. Place a variety of foods out on the playground or in the backyard.

2. Keep an eye open for the different kinds of critters that come to eat.

3. Record on paper or on the tape recorder the children's comments about the different critters and the foods they prefer.

4. Record the times that different critters come to eat.

Want To Do More?

Make a graph (see chapter I on graphing) depicting kinds of animals, frequency of visits and food preferred. Try placing the food out over night. Move the food source to various parts of the yard. Do you think you are meeting the food needs of all animals that might visit? You might consider a hummingbird feeder, or thistle seed for the goldfinches.

KEEP IT IN YOUR MIND AS YOU GO AND FIND

Things You Can Use

area to explore
collecting bags
shoe box

Words You Can Use

collecting
duplicate
find
search
explore
sound
texture
touch
environment

Memory is a key factor in learning. Remembering what you have seen is important in reading and in many other areas. As most children love to go on hunts, this activity is an enjoyable exercise in developing visual memory. In addition, as children learn to sort items by category, they are applying one of the most basic classification skills. The game starts on a simple level and can become as challenging as you and your child want it to be. But, remember, keep it fun!

What To Do

1. Make a collection of items from the backyard or school ground.
2. Place collection in a shoe box—with a lid.
3. Give your child a collecting bag.
4. Open the shoe box and let the child choose two items.
5. Instruct the child to go into the yard to find and collect the same items.
6. The child explores and collects the items in his or her bag, then returns to check the finds.
7. When this becomes easy, show the child the items and have him or her find them without taking the sample along. The adult increases the number of items to be collected as the child's skills increase.

Want To Do More?

Take the items the children have collected and sort them according to similarities and differences.

CREEPY CRAWLER RACE TRACK

Things You Can Use

creepy crawler race track (see illus.)
plastic bags or containers
clear plastic cup

Words You Can Use

fast
slow
start
appropriate insect names

"Oh, daddy long legs with legs so long, you look as though you can move so fast." "Beetle that lies so still, in a flash is past and out of sight." "Ants scurry around and travel routes on stem and ground." But which goes the fastest? Which one won't move when you want it to? Which ones fly away so you can't tell how far they traveled? Let's race to see which is the fastest of the little insects and spiders in your yard.

What To Do

1. Talk about how insects and spiders move, and why. Some animals, like many spiders, sit quietly and wait for food to come to them. Some move quickly to catch food that might get away. Others move quickly so that they can get away! Pictures of insects which show wings and legs will help to illustrate the discussion. Discuss and identify stinging insects, so the children won't try to catch them. (See Bites and Stings, page 00.)

2. Give the children plastic bags or containers. Explain that they are to catch the fastest insect they can find to race against those of their friends.

3. Upon the return of the creepy crawler hunters, begin the races to determine the fastest. To run the race, place the competitors in the center of the race track under a clear plastic cup. Remove the cup. The first to cross the outer most circle wins the heat.

4. After each race, free the losers as close to home as possible. Be careful not to harm the athletes.

5. Keep racing until an overall winner emerges. That's your champion creepy crawler.

6. With much ado, set the champion free to find its mate and have more champions.

Want To Do More?

Discuss what characteristics are necessary for speed. Make a list of the kinds of locomotion: caterpillars, ants, grasshoppers, butterflies, flies, all move differently. In the race time the distances to determine rate of crawl. Or measure how far the contestant will crawl in a given time. Does it move in a straight line or randomly? Vary environmental factors; for instance do they move faster in the sun? the shade? Characterize each creature's movement and try to determine if it is a predator or a plant eater (predators move fast). If you could choose some other way of moving, which would you choose?

SCAVENGER HUNT GRAB BAG LOTTO

Things You Can Use

small paper lunch bags
one large grocery bag
construction paper or cardboard
marking pen
ruler
paste

Words You Can Use

lotto
match
collect
find
search
alike

This is a game that will provide more opportunities for the children to observe, collect, sort and match. It can be adjusted to the age level of any group.

What To Do

1. Take a walk in the backyard, the school yard, the park or the woods.

2. Collect small objects from the environment and place in a collecting bag. Collect objects that are familiar such as pine cones, leaves, twigs, acorns, bark, and rock. Be sure there are duplicates.

3. Return to the classroom or home and make lotto boards. For your board use a piece of construction paper or cardboard 8″ × 12″ for a six picture lotto. Use an 8″ × 8″ piece for a four picture lotto and rule it off into six four inch squares. If you want to have a different color background in each square, you can cut four inch squares of different colored paper and paste one of these in each of the squares on the board.

4. The next step is to sort out the collected items and to place one of each on a master card. This should be done by an adult. Choose items that are small and will fit on the card. After four or six items have been placed on the master card, place the remaining collected objects in a large grab bag. You are now ready to play.

5. Distribute the blank cards to the players, who are seated in a circle.

6. Place the grab bag in the center of the circle.

7. The master card with objects is also in the center of the circle next to the grab bag where it is visible to all participants.

8. The bag is passed around the circle and each participant takes a collected item. If the item is

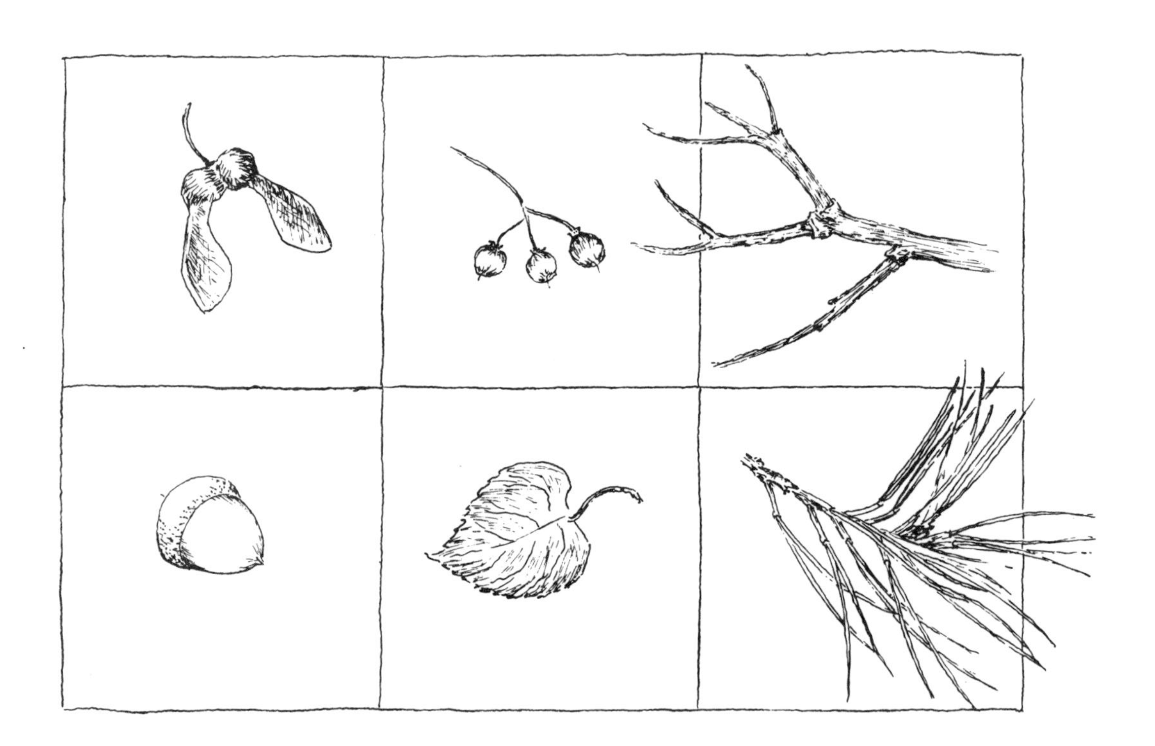

like one on the master card, it is placed in the same position on the individual playing card.

9. The game is completed when someone fills his or her own card with the same objects that are on the master card.

Want To Do More?

An adult draws or pastes pictures of familiar objects on the cards. The cards are then distributed to the players, who then go into the yard to find and collect objects that are depicted on their cards. The game is over when all have found the appropriate objects and returned.

WHAT ABOUT WEEDS?

Things You Can Use

magnifiers
small shovel
scissors
flower pots
writing materials
collecting bags
jump rope or hoop
flower press (See Hodge Podge)

Words You Can Use

weed
tally
census
survivor

Weeds are a nuisance to the gardener, but they can be a great resource for the young scientists. They are plants which can be counted, pulled up, transplanted, dissected, and experimented with, and no one will mind. By taking a plant census, as suggested in this activity, children become aware of the variety and quantity of plants. They also focus on individual plants rather than on the vast mass of green.

What To Do

1. Choose a section of lawn which has a variety of plant forms. Obviously, a lawn well treated with weed killer is not the place to look.

2. Mark off a section, using the jump rope or hoop to form boundaries.

3. Take a census of the plants in the area. Record the varieties you find on a tally sheet (see chapter I). You may also want to write descriptions, use drawings, or press sample leaves. Rank the plants in order of quantity. You may not want to count each grass plant; however, an attempt to do so with older children can give them an idea of the vast number of plants in just one yard, let alone the world. Which are considered weeds and which aren't? Why?

4. Examine your plot immediately following the next mowing. Which plants were damaged the least? The most? Who is the toughest survivor? You might discuss the fact that to grow *just* grass in a lawn requires help—water, fertilizer, weed killer. How could this be related to the fact that grass is usually badly damaged by mowing (i.e almost all its leaves get cut, while weed leaves are below the mowing line)?

Want To Do More?

Transplant some of the plants from one area to another. How does the change affect them? Cover a section of the area with a box for several days. What happens? What happens when you take the box off? Examine and compare the roots, leaves, stems, and flowers of the different plants. Are there things in common? Are there seeds you can collect and grow? Transplant some of the plants into pots and vary the growing conditions (e.g. amount of water, light, soil type). Water with different solutions—salt water, coffee, soapy water, tea, milk, etc. What happens? Is grass ever a weed? (What if it's in the sidewalk cracks or the garden?) Explore a square in another environment such as woods, flower gardens, or marshy areas. Identify your weeds with a field guide.

WHEN IS A GNAT NOT A GNAT? WHEN IT'S A FRUIT FLY!

Things You Can Use

ripe pears
bananas
a jar
cheesecloth
rubber bands

Words You Can Use

fruit fly
adult
larva
pupa
egg
metamorphosis
life cycles

Some animals are so small that they are missed in normal day by day activities. Yet, they are present and can be studied if we take the time to attract them. In this activity we seek out and observe a tiny creature which is found in most environments: the fruit fly. It can be easily attracted with over-ripe fruit. Its life cycle and habits can then be watched and recorded.

What To Do

1. Drop a well ripened fruit into an open jar.

2. Place it outdoors in an area out of the sun where flies and other insects will have easy access.

3. Watch daily for the fruit to attract the tiny flies.

4. If you see the tiny fruit flies, you will know that their eggs have been deposited. Close the opening of the jar with cheesecloth fastened with a rubber band.

5. The tiny eggs (microscopic) will hatch and the larva will be seen on the fruit and jar.

6. The complete life cycle will be carried out in this jar. The larva will pupate and finally turn into an adult fly. This adult will then mate and the cycle will repeat. The whole cycle takes ten days to two weeks.

Want To Do More?

Traps can be set out for other insects. Raw meat will attract other flies and some beetles. Sugar will attract bees, ants and wasps.

You will find that some creatures can be caught during the day and others only at night. The fruit fly is a pest. What other insects are pests? Why are they important in nature?

I CAN MAKE WHAT YOU MAKE

Things You Can Use

objects collected on a field trip (this activity requires a matched set of five to six pairs of items)
tag board or construction paper—one piece for each child, and one for the adult

Words You Can Use

communicate
top
bottom
middle

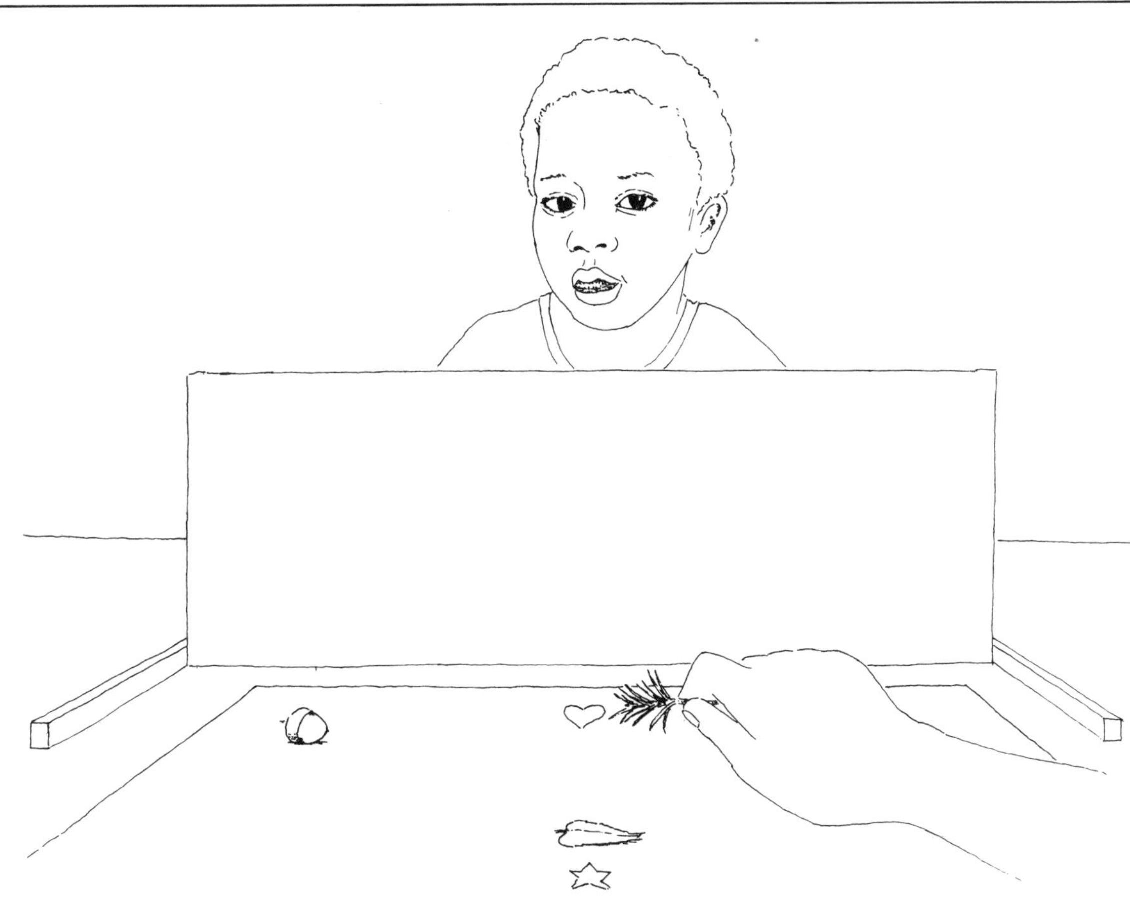

An essential skill in our society is effective communication; this involves both speaking and listening. "I can make what you make" requires accuracy in both. By giving and following instructions carefully, partners work together to create identical pictures and to improve their communication skills.

What To Do

1. Place a paper rectangle in front of each child.
2. The adult places three to six items on his or her paper.
3. The adult then tells the children where to place the items on their papers.
4. After finishing, compare how the items were placed.
5. If problems result, place some landmarks on the child's paper. Use these to locate things (i.e. a star at the top, a heart at the bottom, "put the leaf at the top, next to the star").
6. When the children become adept at following the adult directions, pair them off and have one partner tell the other what to do.

Want To Do More?

Add more objects. If younger students are having difficulty, allow them to place the objects while looking at the other paper.

An exercise which increases visual memory is to give the children a few minutes to look at the teacher's paper and to memorize the location of the objects. Then they place them on their own papers, from memory.

THE SUN WON'T SHINE THROUGH YOU

Things You Can Use

materials which have the physical properties below, i.e. clear glass, soda bottles of various colors, cloudy glass, black paper, etc.

Words You Can Use

transparent
translucent
opaque

You have sunshine and a nice warm day. What can you do with them that will be both fun and enlightening? Explore the power of the sun. What will it go through and what blocks it? Just like Superman's X-ray vision, it won't shine through everything, but it will go through a lot. In this activity, we explore different materials and the light from the sun. Note: looking directly into the sun can cause *serious* eye damage. Be sure to caution the children against it.

What To Do

1. Select materials which have the properties above. Present these to the children. Let them discover by observation which materials the light will shine through.

2. Have the children look through these same objects. Some they will be able to see through, some they won't.

3. Now mix all the objects so that the students can sort them.

4. Have the children sort the objects first by whether they think they will be able to see through them.

5. Have them check the results by actually looking through the objects.

Want To Do More?

Discuss reflecting and absorbing powers. Play with mirrors and light.

TOSS AND FIND

Things You Can Use

chart of attributes (see illustrations)
crayons
coin or marker

Words You Can Use

attributes
traits
common
collect
coordinates
characteristic

One of the most direct ways for children to organize their impressions of the environment is to classify collected items according to common attributes or traits. When they are looking at living plants, the children can see how plants are alike and how they differ. For example, some may have yellow flowers, some, smooth leaves; some may grow as vines. Often they can be grouped by one common characteristic. In some cases, the same plant may be placed in several groups. The structure of this activity introduces the child to the concept of reversibility: moving from the general to the specific and back to the general. The concept of reversibility is essential to logical thinking and needs much reinforcement in the early years.

What To Do

1. Construct a chart similar to that in the drawing. This chart indicates plant possibilities that one could collect in a given area. It may include an impossible combination. It's fun to watch the kids when they realize this.
2. Instruct the children in the use of the chart.
3. Move to the chosen collecting area.
4. Select a coin, place the chart flat on the ground and toss the coin to land on the chart. Locate the site of the landing and read the coordinates.
5. Go out and find a plant sample to meet the requirements of the coordinates.
6. The rest of the class checks the plants as they come in and colors in the space when the requirements have been met.

Want To Do More?

Try this chart with any set of natural objects, i.e. sea shells on the beach, rocks in a gravel pile or rock collections, insects.

Make lists of things you are unable to find. Why? Where might you go for these?

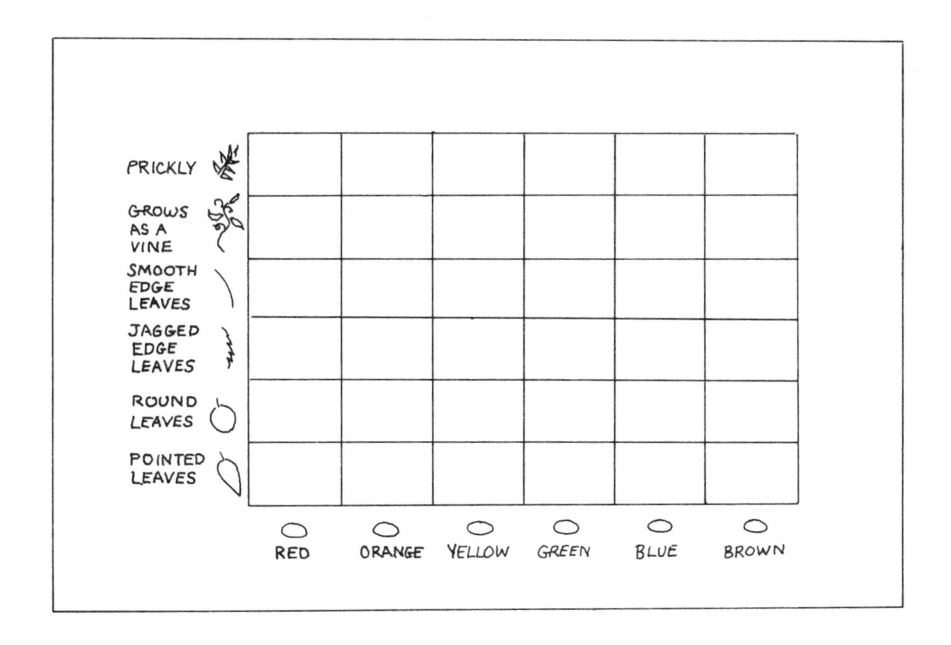

COUNT WHAT YOU CAN AND DO IT AGAIN

Things You Can Use

leaves
bushes
plants
paper
pencils

Words You Can Use

count
classify
color
set
properties
total
number

Mathematical skills and concepts are developed through use. This activity provides practice with the basic mathematical operation of counting. It capitalizes on the stage most children go through when they first learn to count: wanting to count everything. Using the outdoors provides new materials to count, as well as giving the children a chance to explore. Note: adjust the number of leaves to meet the children's counting abilities.

What To Do

1. Take the children outdoors into the backyard or the school yard.
2. Collect one leaf from each tree in the yard.
3. Count these leaves to find the total number.
4. Classify the leaves on the basis of color, placing all the leaves of the same color in the same group.
5. Count the number of groups.
6. Count the number of leaves in each group.

Want To Do More?

Classify the leaves on the basis of other properties (such as shape) and then count the groups, and the members of each group. Repeat the above activities with leaves from bushes and plants.

WANT TO KNOW HOW MUCH?

Measurement Experiences

THE BIG SQUIRT

Things You Can Use

plastic squeeze bottles

Words You Can Use

water
full
squirt
how far
farther
farthest

Kids love water! They like to run through it, wade in it, pour it, and squirt it. This activity lets them have fun with water as they explore concepts of far, farther, and distance. Choose a nice day.

What To Do

1. Fill plastic squeeze bottles with water.
2. Make a line on the sidewalk or stand at its edge.
3. Let the children squeeze their bottles to see who can squirt water the farthest.

Want To Do More?

Let the children draw water pictures on the sidewalk, talk about evaporation, let them squirt each other.

Try it with different sized bottles. What designs do the squirts make? Who can empty his or her bottle fastest?

MEASURE THE WIND!

Things You Can Use

the wind
dandelions

Words You Can Use

wind
hard
soft
directions
N, S, W, E
velocity

The wind blows everyday. Sometimes it blows hard—sometimes softly. We can feel it blowing through our hair and around our bodies. Kids can blow like the wind, too. They do this when they blow out the candles on a birthday cake. When they blow softly, the candles stay lit. Then they blow hard—out they go! Kids also like to blow dandelions in the spring. This activity lets them blow to their hearts' content as they learn about wind velocity.

What To Do

1. Talk about wind as moving air. Try blowing hard, then softly.
2. Go outdoors and select a mature dandelion. Blow this dandelion to demonstrate hard and soft blowing.
3. Now take a dandelion and shake loose the seeds so they float in the wind.
4. Notice the direction the wind moves them.
5. Discuss how the wind blows both hard and soft and how in each area, it usually blows in certain directions.

Want To Do More?

Draw a series of circles in the school yard. Use these circles to measure distance and draw lines to show flight direction. Determine the strength of the wind and the direction of the seeds' flight. Discuss seed dispersal by wind. Develop wind vanes and other wind direction devices. Can you find examples of how wind has affected your environment (i.e. plant growth, land formations)?

HOW DO YOU MEASURE THINGS?

Things You Can Use

balance
containers
sticks of varying lengths

Words You Can Use

measure
weight
length
volume
how much
how heavy
how full
how long

Children see their parents taking measurements. They see them step on the bathroom scales, measure a board for a shelf, weigh bananas at the store. They watch the numbers go around on the gas pump. They get weighed and measured at the doctor's office. Children know all this is important, and they will imitate measuring in their play. They hop on the scales and say, "Oh no! Fifty pounds." They fold and unfold a carpenter's rule and tell you the table is "twenty-three foot inches big." Measurement plays an important part in all our lives. This activity introduces children to the basic kinds of measurement and gives them a little insight into the activities of the bigger people in their world.

What To Do

1. Introduce measuring as something people do when they want to know how much of something they need or have.

2. State that people usually find out how much in three ways: how long? how heavy? how full?

3. Each child will need one object.

4. The child will measure how long it is by comparing it (longer or shorter than) to a given length (i.e. foot length, book length, paper clip length,etc.).

5. The child will compare the weight of the object with a standard unit on a balance. The standard unit can be a rock, a can of sand, a babyfood jar or anything else. Is it lighter or heavier or the same?

6. The child will find out how much space the object occupies by placing it in a container such as a glass, a souffle cup or a babyfood jar. The object will fill up the jar, not fill it up, or not fit in (meaning it fills the container too full).

Want To Do More?

Find a partner and compare objects with your partner. Introduce the standard units of measurement. Take a measuring field trip to the supermarket. Use each measuring unit above to measure lots of objects.

HEAD TO TOE ALL IN A ROW

Things You Can Use

kids and other miscellaneous materials

Words You Can Use

measure
big
bigger
biggest
and other appropriate comparative terms

Although meters and miles are meaningless words to young children, they are beginning to understand measurement. They know that one friend lives next door while another lives "way far." They know the little slide is exciting, but the big slide is just a little too big to be fun. Brooke is smaller than Felice, and Alicia is the tallest one in the class. They especially know, "Carlton got the big piece and I only got the littlest one." You can capitalize on this interest in comparisons by measuring with things familiar to the children. Even children who can't count are capable of comparative measurement. Once started, they want to measure everything.

What To Do

1. Begin with some "which is bigger? (or longer or smaller)" questions. Compare things which can be placed side by side. Sticks, toys, leaves, kids, dogs, or grown-ups can all be used. Children need to be successful in this activity before moving to the next.

2. Measure some of the same objects using something such as blocks. Discussion will bring out comments like "It took a lot of blocks for Mrs. G. It was only a little pile for Anna." or "The guinea pig cage was a long row. The mice was short." Writing such comments down and reading them back can reinforce the experience and it also helps with the time concepts ("Remember what we did yesterday?").

3. The children are now ready for some big measuring. Start with an "I wonder" question, "I wonder which is longest, the front sidewalk or the back?" Choose things with fairly different dimensions. Use the kids to measure. Place the children, lying down, head to toe first along one sidewalk, then along the other. Comparing, rather than counting, ought to be stressed. Children may remark that "John and Rachel got to lie down on the front sidewalk, but there wasn't room on the back one." Such comments should be encouraged by asking "Why?" and "What else did you see? Do remember, though, that repeated experiences often teach more than one experience "over-talked." Many children who enjoy this activity may not grasp the concept until they have experienced it several times in varied ways.

Want To Do More?

Find a large tree. How many kids does it take to hold hands around it? How many shoes long is the couch? Can you find something the same size as your foot? Your arm? You? Trace the children and cut the figures out. Measure with the paper people. Make a paper chain the height of each child in the class. Connect them. That's how tall the class is. Use it to find out how many kids it would take to reach the ceiling or the tree branch. With children who can count, count how many steps it takes to walk across the yard.

THE LONG AND SHORT OF IT

Things You Can Use

ten twigs, cut into various lengths

Words You Can Use

short
long
small
large
shorter
longer
smaller
larger

This more complicated activity moves the children beyond comparing two objects to ordering a whole group of objects according to size. They then begin to realize that long, short, and other comparative terms are not fixed labels, but vary with the situation. It is also a good way to use all those sticks kids love to pick up.

What To Do

1. Ask children to arrange the precut twigs in sequence by length from shortest to longest. Begin with three to five twigs of distinctly different sizes.

2. Gradually add more and refine the differences.

3. Repeat the above, but reverse the sequence from longest to shortest. Assist the child as necessary—but avoid doing the activity for him. Ask questions to get the child to observe an error if one is made.

Want To Do More?

Repeat above procedures using other objects, for variety (e.g. pipe cleaners, paper towel tubes, sticks, leaves, etc.).

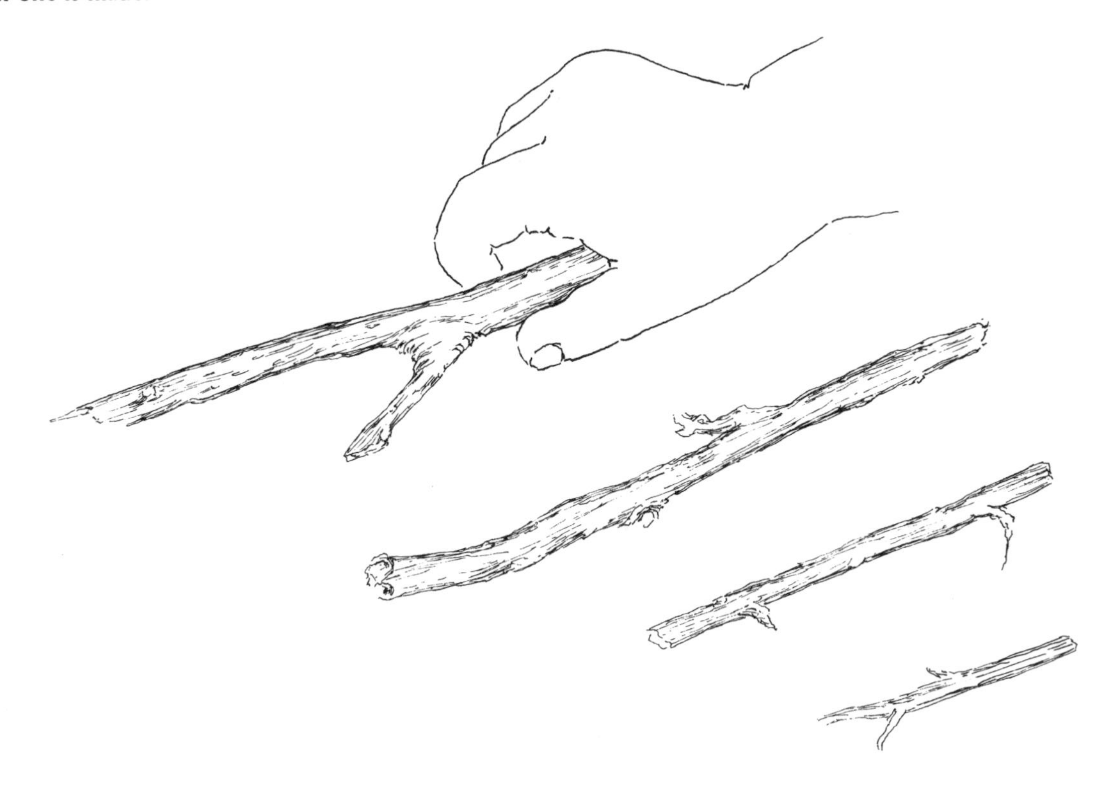

SNOWJOB

Things You Can Use

clear plastic glasses
snow
masking tape
markers

Words You Can Use

snow
ice
water
freeze
evaporate
melt
more
less

Kids love snow and are eager to participate in any activity involving it. In this activity we explore the physical states of water. Understanding these processes is beyond young children, but exposure to them provides a basis for later comprehension. It also encourages close observation and real thinking. As a bonus, it can provide some entertaining conversation for the adults involved. Please don't laugh, though; the children are serious.

What To Do

1. After a snowfall, have the children fill a plastic glass to the brim with snow. Bring it inside.

2. Ask the children what they think will happen to the snow. Expect a variety of answers. "It will turn into ice." "It will melt." "Nothing." "It will be water." Don't choose a correct answer, just encourage speculation.

3. As the snow begins to melt, ask the children what is happening. "It's turning to water," "It's getting little," and "It's going away" are typical answers. Look for observant remarks, but accept all comments.

4. Now ask why the snow is melting. If the kids feel free to think and speculate, the answers can become really entertaining.

"It must be the sun."
"Well, let's pull down the shades then."
"It's still melting."
"Let's turn off the lights and close the door. Maybe that'll work."
"Oh no! It's still getting to water!"

5. After much speculation and discussion, the adult may want to talk about temperature indoors and out, relating it to clothes worn, how hands feel, or other familiar events.

6. Once the snow is melted, mark the much lower water level. The reason for the lower level can be demonstrated simply. Show the children a cupful of crumpled pieces of paper. Flatten the pieces out and show how much less space is used. They will not understand completely, but it does give them something to think over.

7. Next, ask the children what they think will happen when the water is put back outside. Some expect it to change back to snow, others will say ice. Do it and watch what happens. Again, don't explain, just observe.

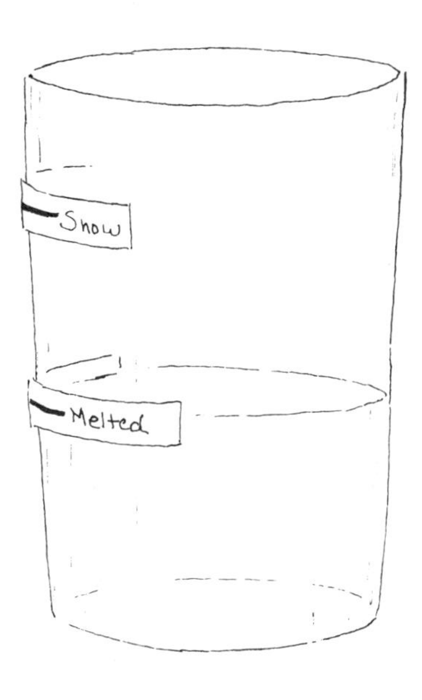

8. Finally, bring the ice in and let it melt and evaporate. Although the children will not understand the process, it is interesting for them to observe and to mark the changing water level. Look upon these initial experiences as a chance to observe and to talk about the observations. If you really want some interesting comments, just ask why the snow is white and water is clear!

Want To Do More?

Melt snow in different locations and compare rates. What are the effects of salt? What happens when salt water evaporates?

LONG, SHORT, TALL, SMALL— COMPARE THEM ALL

Things You Can Use

pine cones (big and little sizes)
flowers/stems (tall and short)
leaves (big and little)
animal tracks (big and little)
earth worms (big and little)
seed pods (long and short)
twigs (long and short)
pictures of animals (big and bigger)
ants (small and smaller)
building blocks (stack tall, stack short)
kids (stand back to back, who's the tallest? the shortest?)

Long before children learn to measure in standard units like meters or pounds, they begin to compare. These comparisons are made with things they can see or handle. In this activity, children compare like objects directly in terms of size. They learn that things can be the same in some ways, yet different in others, a concept they need to explore again and again.

What To Do

1. Lay out two of the objects listed that have the same shapes but different sizes. Ask the child to touch the big one—then the little one.

2. Follow the same procedure with appropriate variation, for each of the items that are listed in "Things You Can Use."

3. The children should be encouraged to pick up and handle the items in order to physically compare them. As they touch, they should be encouraged to talk about the object's size in comparison to another like object.

Want To Do More?

Do leaf rubbings on a nature hike. Compare the sizes of the rubbings. Sort according to size. Take a walk. Look at trees. Find the tallest tree, the shortest tree.

TIN CAN ICE CREAM

Things You Can Use

soda pop or beer cans (wash thoroughly and remove the tops leaving no sharp edges)
snow or crushed ice
tongue depressors or popsicle sticks
milk
flavoring
plastic cottage cheese or liver containers from the supermarket
sugar
canned milk
spoons

Words You Can Use

freeze
mixture
mix

Ice cream! Ice cream! We all love ice cream. It's even better when you make it yourself, especially if you can taste it as you go along to be sure it's just right. Although this activity requires a little advance preparation by an adult, once everything is set up, it's easy to do. Can you think of a better use for the first snow of winter than to use it to freeze ice cream?

What To Do

1. Select a can for each child. Aluminum cans let the ice cream freeze faster, but soup cans will work, too.

2. Select a plastic container for each child. Place some salt in the bottom of it.

3. Place the cans in the plastic containers.

4. Layer snow and more salt into the plastic containers until the snow is high on the sides of the cans.

5. Pour 150 ml (½c) of milk into each can.

6. Add 15 ml (1 T) each of sugar and canned milk.

7. If flavoring is desired, add your favorite, i.e. vanilla or other extracts, chocolate.

8. Stir the mixture with the wooden tongue depressor. The key here is to scrape the freezing ice cream away from the sides of the can so that more milk can be frozen.

9. When the mixture has reached the consistency of a thick shake, it's ready to eat. This takes about 10 minutes. Add more salt to the snow if it's taking too long. Tasting makes the time go quickly!

Want To Do More?

Add fruit and nuts to the mixture. Use thermometers to check the temperature of the snow, the snow and salt, and the milk mixture. Measure the amount of salt added to the mixture to change freezing time. For those who can't have milk, freeze juices. Another approach is to place a small covered jar of the ice cream mixture in the middle of a coffee can filled with snow and salt. Put a lid on the can. shake the can until the ice cream is frozen.

HOW DOES YOUR DANDELION GROW?

Things You Can Use

stick
crayon or magic markers
dandelions

Words You Can Use

grow
dandelion
plant
tall
compare

City child or country child, all have seen a dandelion. Barefoot babies scooting through the grass their first summer will squirm until they finally reach it. Many parents have been presented with a pretty yellow bouquet collected by small hands. Do young children realize that the yellow flower turns into the seed puffs they blow away? Most do not. This activity focuses their attention on the changes in this rapidly growing plant and allows them to see a life cycle from "start to finish." If you want to go all out, you can even plant the seeds and grow some more!

What To Do

1. Find a young dandelion plant.
2. Insert a stick into the ground next to the plant.
3. Each day as the dandelion grows taller make a mark on the stick. See how tall it grows in one week, two weeks.
4. Discuss what the dandelion needs in order to grow. Compare to what children need to grow.

Want To Do More?

Repeat above procedure using more than one dandelion. Compare to see which dandelions are growing faster. Compare some that are in the shade most of the day with those that get lots of sunlight. Discuss the differences.

IT TAKES A WHOLE BUNCH

Things You Can Use

containers
objects of varying sizes collectible in large quantities (e.g. acorns, pine cones, sweet gum balls, rocks)

Words You Can Use

big
little
small
much
many
a lot
a few

Children work with volume every day. They pour milk and judge when to stop. They usually know if a friend got more. They play with sand and water, pouring from one container to another, and, eventually, through experience, learn to avoid overflows. In most instances, however, these experiences involve comparisons of like substances.

In this activity we explore volume through comparing unlike objects. This helps children to understand that equal volume does not necessarily mean equal quantity.

What To Do

1. Set several containers out in the yard, using two of each kind.

2. Discuss amounts such as many, a lot of, much, few, big and little.

3. Select two same sized containers to fill with your chosen objects. One can might be filled with pine cones, the other with acorns. Be sure to have a significant difference in the sizes of the objects.

4. Note the difference in the time required to fill each container.

5. After they are filled, dump out the contents and discuss the two different piles.

Want To Do More?

Count the objects in the containers. Record the number required to fill each container. Write the numbers to show that the largest number means the most objects. Compare acorns to pine cones by first making a row of pine cones, then placing one acorn next to each pine cone. Continue making rows of acorns until all are used. It becomes obvious even without counting that there are lots more acorns than pine cones. Children with number skills can determine the ratio of acorns to pine cones.

Bring in a bigger container. Guess how many will be needed to fill this one. Find bigger objects to use.

MEASURE SHADOWS

Things You Can Use

permanent objects in the yard or on the playground, i.e. garbage can, swing set, clothes line pole, climbing frame, flag pole, tree, bush
stakes with ribbons colored differently for each season

Words You Can Use

winter
spring
summer
fall
seasons
big
bigger
small
smaller
size
shadow

Shadows come in all shapes and sizes. Even a single object produces a shadow which varies with the time of day and the seasons. In this activity children can observe and record changes in shadows from season to season. The helping adults may learn a little, as well. After all, do you know how the shadow of your tree changes through the year?

What To Do

1. Inside or outside, discuss shadows.
2. Make some with your hands and with your bodies.
3. Go outside and look for shadows. Discuss how large the shadow is.
4. Measure it and mark it with the colored stakes.
5. Repeat the activity at different times of the day. Are the shadows longer, or shorter? Why?
6. Repeat the activity as the seasons pass.
7. Photograph at least one shadow each season. Do the shadows remain the same size? Are they different? Why?

Want To Do More?

Repeat the above procedure, only this time use chalk and trace around the shadows. Do this at various times of the day and during different seasons. Compare your various tracings.

Using a strong flashlight make a shadow of your hand on the wall. Next, use items from outside—i.e. leaves, pine cones, branches, plants and let the children identify what the shadows are.

FILL THAT SPACE— AN AREA GAME

Things You Can Use

cards cut from tagboard in various sizes. The cards should vary in size, and with more skilled children vary in shape.
leaves

Words You Can Use

leaves
same
compare
size

Even young children have experienced the concept of area. They learn to assemble simple puzzles. They fit blocks together to cover spaces. They spread blanket on blanket until the floor of a "hideout" is covered. As with the other measuring activities in this book, the children are asked here to compare and match rather than to use standard measurements. In this case, they are off on a leaf hunt, looking for the leaf which is "just almost the same" as their card in size.

What To Do

1. The objective of this game is to find leaves from the outdoors that will fill or cover the shapes cut from the tagboard.

2. Explain that the purpose is to "just about" cover the card. You will want to construct the cards based on normal sizes of leaves in the yard.

3. The cards may be placed in sets or given out individually.

4. Move outdoors and collect a pile of leaves from various trees. Place the cards on the ground and go through the leaves with the children, searching for leaves which correspond in size to the cards.

Want To Do More?

Count the leaves that are the same size. Time how fast the leaves can be found. Order leaves and cards by size. Add shapes, circles, and triangles. Use a scavenger hunt of shapes and different sized shapes. Graph the results.

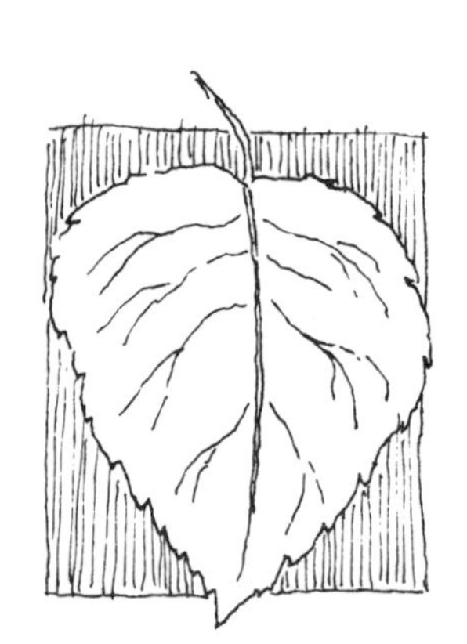

NUTS FOR YOU AND ROCKS, TOO

Things You Can Use

balance scale
rocks
buckeyes
acorns
eucalyptus buttons
etc.

Words You Can Use

balance
scale
heavier
lighter
and other comparative terms
weigh
compare

Give children a balance scale and a bucket of junk and in just a few minutes they will be totally absorbed in experimenting. An adult doesn't need to utter a word of explanatory "how to use its." Given the freedom to explore, children will naturally begin to work toward "making it straight" and will investigate many other problems as well. Only after they've had the opportunity to work freely for a period of time is it appropriate for the adult to go beyond commenting on and discussing the children's work. When the initial interest seems to be waning slightly, the children are ready for something new.

What To Do

1. You want to give the children the idea of comparing the weight of several things to one. Choose something to be the constant. It could be a rock, a cup of acorns, three walnuts, whatever. Place it in one pan of the balance.

2. Now it's time to compare. For example: "I put the red cup of buckeyes on the balance. What do you think will happen if I fill the other red cup with rocks and put it on the other side?" Or, "What can you find that will make the rock go up?" Or, "What can you put on the other side that will make the balance straight?" Children may say something like "I guessed the big pile of buckeyes would make the walnuts go up, but they didn't." Encourage comparative language in whatever form the children use it.

3. Once the children are familiar with this, play "collect and guess." Have each child collect something from outside (i.e. a rock, some pine cones, nuts). Choose something, such as a rock or a can full of acorns as a constant. Let everyone hold it, then put it on the balance. Each person is to guess whether their object(s) is heavier, lighter, or about the same as the constant. An adult writes down the guesses. Each person in turn places his or her object(s) on the scale to find out how accurate the guess was. You will find that some children become very accurate, while others continue to base their guess entirely on the size or volume. This is a valuable activity for both groups to experience repeatedly. And they all enjoy the guessing!

Want To Do More?

With older children, you may want to introduce standard units of measure such as grams and ounces. With children who can count, numerical comparisons can be made (i.e. 3 rocks balance 9 pecans). Compare the uniform weights of man-made objects with the uneven weights of natural objects. The same size of washer will always weigh the same, whereas each acorn will be a little different. Compare the same quantity of different substances—3 peach seeds with 3 shells or a cottage cheese carton of styrofoam with one filled with corn.

HOW DEEP IS YOUR SNOW DRIFT?

Things You Can Use

stick
permanent marker
calendar

Words You Can Use

snow
freezing
melt

The first snow of winter is special. So is the first time it is deep enough for us to go out and sled. Especially in warmer climates, snow is a treat—and fun! But as the winter passes, we do not notice the changes in the snow until it melts enough for the ground to show and the sleds to start dragging. The purpose of this activity is to observe the depth of the snow over a period of time, and thereby to recognize the effects of different kinds of weather, such as, cold and warm spells, wind, and additional snow storms. It is a simple way of recognizing changes which often go unnoticed.

What To Do

1. Before the first snow, place sticks in spots where they can be easily observed and will not be in the way.
2. After the first snow, go out and mark the depth of the snow and the date on the stick, using the permanent marker.
3. After a few days or hours, depending on the weather, measure and mark the depth again.
4. After each temperature change or snow storm measure and mark the depth and date it.
5. When the snow has gone, note the different depths that you have marked for your snow field.

Want To Do More?

Make a paper snowflake (see Hodge Podge). Compare snowdrifts in various spots, for instance a windy spot and a quiet spot. Do your measuring with a meter stick. Discuss how snow forms. Ten inches of snow equal one inch of water. How much moisture did your snow provide? Melt snow. Find temperatures on various days at various spots. Record weather conditions on a calendar, along with changes in snow depth. Observe weather conditions and try to predict the effects on the depth of the snow.

WATCHING TIME & SEASONS

Change and Time Experiences

BURY THE SOCK

Things You Can Use

nylon sock
leaves
plants
grass
tin cans
glass
plastic
tissue
cellophane wrapper
gum
apple core
fast food containers

Words You Can Use

collect
litter
bury
change
time

We often hear that litter isn't biodegradable. Does this mean anything to a young child? Probably not. To an adult, it means that most litter does not break down biologically. What the child needs to know is that litter doesn't rot or decay very easily; it just stays where we leave it. In this activity the children discover what happens to cast-offs from people and from nature.

What To Do

1. Take a nylon sock and collect both natural materials and litter that you find in the yard. Try to get at least one each of the items listed in "Things You Can Use."

2. Make a record of your collecting by sketching the sock and its content.

3. Bury the sock so that it is completely covered by dirt.

4. Dig up the sock in a few months.

5. Compare what has happened to the organic materials with what has happened to the glass, plastic and metal. Talk about it. You can even moralize a little if you like. The idea that trash dropped on the ground doesn't just disappear is important—even for young children. Attitudes toward the world form easily. They ought to be good ones!

Want To Do More?

Keep the sock in the ground for one year. Repeat the procedure outlined and discuss changes.

THE SEASONS GO ROUND AND ROUND

Things You Can Use

camera
string
paper clips
tape
hula hoop

Words You Can Use

photograph
picture
time
change
season
events
grow
winter
spring
summer
fall

By looking at photographs and listening to family lore, young children can grasp the idea of their own growth from babyhood to childhood. By the age of three, they, themselves, can remember some version of "last year." They see snow and recall a snowman, "we made that time." They talk of camping, "when the warm gets here," or they say that, "we're going to have Easter bunnies when there's grass." Through the use of photographs of themselves and their friends in seasonal settings, these beginning concepts of time can be expanded. They can begin to understand the predictable cycle of the seasons, which contains the continuous, linear nature of their own growth. Consistent patterns give order to all life. Within them, there is the infinite variety which makes life so interesting. There will always be another winter, but there's only one winter when you're three.

What To Do

1. Collect photographs of children outside during the various seasons. Photos should be of activities children might engage in during the particular season. (Fall—raking leaves, winter—wearing warm clothes, summer—running, spring—planting flowers, etc.) Families often have such photographs in their photo albums. Teachers can take photos through the year.

2. Let the children pick photos that depict the four seasons.

3. Attach the photos to a hula hoop with string and paper clips, ordering them in the proper time sequence. (i.e. Fall, Winter, Spring, Summer.)

4. Hang the hoop at eye level by suspending it from the ceiling with at least three strings.

5. Discuss the events, time sequence, and changes with the children.

Want To Do More?

Cut pictures from magazines depicting various seasons. Make a booklet of pictures for each season.

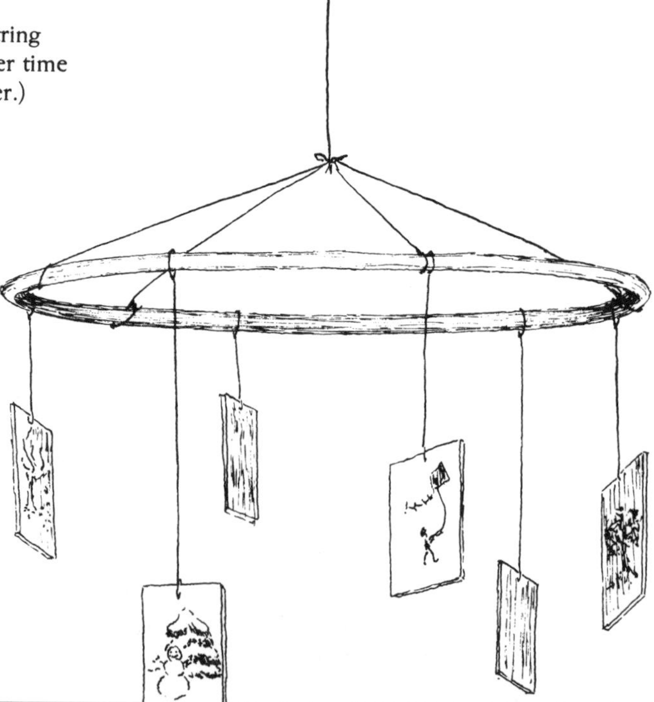

QUIET TIME

A quiet time is a time for rest, a time to think, a time to wonder, a time to plan, a time to dream. Real quiet time is rare for many children. Much of the time they spend alone they spend watching televison or playing with toys. While these activities should not be eliminated, children do need to learn about quiet time alone to think and dream. By spending time outdoors, children can learn to study the environment through unhurried, careful eyes. In this activity, we take a normal rest period for young children and move it outside, giving them the opportunity to "just sit quietly and smell the flowers."

What To Do

1. When warm weather arrives, choose a quiet spot suitable for moving your rest period outdoors. It is preferable to choose an area that the children do not normally use for play. Try the front yard.

2. Provide each child with something to lie on, preferably, their regular sleeping mat. Stress that this is still rest time, a time to be quiet, not to play. The teacher should have a space, too. Place the kids far enough apart to give each person an individual space.

3. When rest time is over, talk about the new experience. What did you see, hear, and smell? How did the air, wind, and sun feel?

4. Continue this for at least a week. This allows time for the novelty to wear off. The children can then settle in and relax with their outdoor quiet time.

Want To Do More?

Set the mood and perhaps a focus of discussion by reading a story or some poems about trees, grass, clouds, or other suitable topics.

SPRING IS THE TIME FOR BABY BIRDS

Things You Can Use

assorted pieces of string
cotton
dry grass
yarn—provide a variety of colors.
a mesh onion bag

Words You Can Use

nest
spring
mate
male
female
bird names as determined by locality

Spring is a time for birds, for robins returning, for the sounds of mother and father birds finding their nesting sites, beginning the ritual of nest building again. What do birds use for nesting materials? Do they prefer certain things above others? In this activity we focus on birds and their nests. The children learn which materials certain birds use, and they help these birds as they go about the business of preparing for their babies.

What To Do

1. Cut up pieces of string, yarn fuzz, clothing strips, etc. Use 5–6 inch pieces. Have the children save these to bring to the birds.

2. Place the collection in a loose sack so birds can remove them. An onion or fruit sack of loose plastic weave is excellent.

3. Place the sack in a spot where you can observe it while not disturbing the birds. This should be done in March.

4. Watch the birds as they visit the string sack.

5. Discuss the choices. Are there favorites? Is there something no bird wants?

Want To Do More?

Do the father and mother bird work together to build the nest? If you can, watch nest building from start to finish. How long does it take? Why do you think the birds built their nest in that special place? Is it an old nest or a new nest?

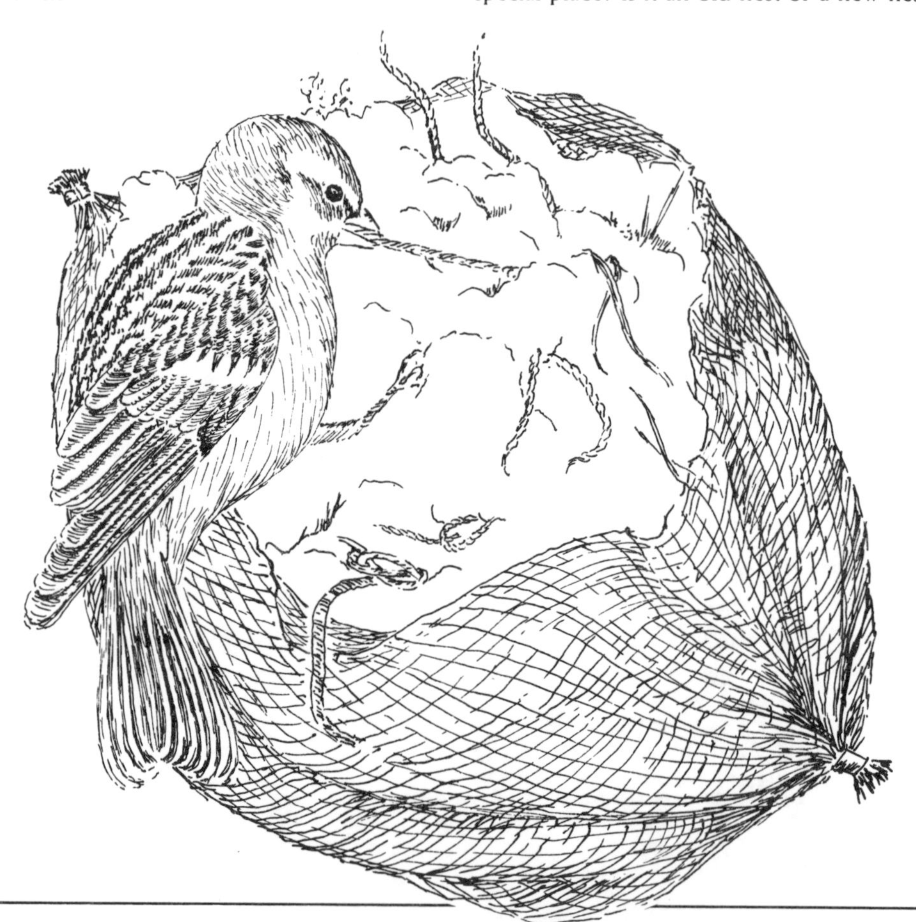

WHAT COLOR IS SPRING?

Things You Can Use

box of crayons
tagboard divided into columns with one each of the following colors at the top of each column: red, orange, yellow, green, blue, violet, brown and black; keep one column open for colors that do not match those presented

Words You Can Use

names of colors
collect
mark
seasons

Spring is a wonderful annual event. It is a time of new life, growth, sweet smells, warmth, and love. There are many new colors that suddenly appear following the bitter cold and dreariness of a long winter. These are the colors of spring. This activity helps the children develop an awareness of the sights and colors of spring and lets them use their observing and recording skills.

What To Do

1. Make a color tally card for each child. Tagboard should be divided into eight columns (see illustration). Each column is for recording one color. Children can help make the tally cards by coloring the top of each column with a different color.

2. Discuss spring. When does it come? What changes take place? What new sounds do we hear? What new colors do we see? Photographs, as well as real objects such as grass or a daffodil may be helpful.

3. Take a walk in the spring. Let each child take a color tally card. Each time the child sees a specific color he or she makes a mark in the column under that color.

4. Upon return to the home or classroom, the children will have a record of the colors of spring that they have observed.

5. Discuss the colors. Compare the tally cards; did everyone find the same colors?

Want To Do More?

Take a color walk during each of the seasons. Check the colors of the seasons against each other. What color differences do you notice? Why do colors change from season to season? Which season has the most colors? Which season has the least colors? Which season has the most yellow, the most blue, etc.? Using pictures from magazines or collections of sticks, grass, leaves, buds, etc., make a collage for each season. Keep and compare.

RED	ORANGE	YELLOW	GREEN	BLUE	VIOLET	BROWN	BLACK	

OLD AND NEW

Things You Can Use

items collected in the back yard or school yard, i.e. leaves, twigs, grass, plants
paper lunch sacks

Words You Can Use

time
old
new
young
fresh
dry

How can a young child contemplate those things that are a million years old or older? We adults find the idea of old things fascinating, but our imaginings are based on a well developed concept of old. For a child, old is yesterday, and time has just begun. If a pair of tennis shoes is old because it was purchased last year when the child was two, how can the child imagine a tree's age when it has existed forty times longer than his or her total life span? Although it is impossible for three year olds to understand "a million years ago," they can begin to develop a concept of old and new which can be refined through experiences such as this one.

What To Do

1. Go into the back yard or school yard with the children. Each child should have a paper collecting bag.

2. Tell the children that they should look for and collect things in the yard that they feel are old as well as things that they think are new.

3. After collection is completed—5 to 10 minutes—have the children gather in one place. Take items from the collecting sacks.

4. As you discuss each item, ask the children to tell you whether they feel a particular item is old or new. New might include leaves, grass, flowers. Old might include rocks, sticks, dirt. The oldness or newness of a collected item depends on its relationship with the rest of the collection. For example a rock is old compared to a tree, a tree is old compared to a newly sprouted plant. A wilted flower is old compared to a fresh blossom.

Want To Do More?

Walk around and find things that are old and used and new. Discuss the differences between these concepts.

WASH OUT

Things You Can Use

a sprinkling can
two quarts of moist soil

Words You Can Use

pour
run
washed down
gully
erosion

Erosion carves out canyons, erases mountains, creates river channels, and shapes land masses. Although young children cannot comprehend these massive efforts, they can see gullies form on a grassless hillside. A trip to a construction site will clearly show what can happen to unprotected soil. This activity introduces the word erosion. Through a simple demonstration, children become aware of the effects of water.

What To Do

1. Take the children on a walk or drive through a recently constructed housing subdivision.
2. Look for signs of soil erosion.
3. Examine the erosion closely. Discuss how the soil can change and what things can change it, i.e. water, wind, weathering.
4. While at the construction site, gather the children around you and make a pile of dirt on the ground. Let the children imagine that the mound of dirt represents a real hill.
5. Pour water slowly from a sprinkling can over the mound of soil.
6. The children will see how the soil is washed down the hill and how the hill is gradually leveled.
7. Where does the soil go?
8. What makes the gullies?

Want To Do More?

Take a walk in your neighborhood or in the park. Look for signs of erosion.

On a rainy day collect a glass of water from a gutter. Take it in the house, examine closely—see the soil in the water. How did it get there?

Discuss what helps stop erosion (i.e. grass, trees, leaf mulch, straw).

Draw a picture of what your neighborhood would be like if lots of soil washed away.

HAPPENINGS DIARY

Things You Can Use

scrapbook
diary (ready made)
looseleaf notebook
or your own book made of heavy construction paper or drawing paper
paper punch
yarn
string
brads
photos
children's drawings
pictures labeled by adults
crayons
paste

Baby books record the significant events of a child's first years. Photograph albums or vacation scrapbooks record family happenings. The Happenings Diary is another record that helps a child remember special occurrences in the natural world, as well as those at home or school. It can be as much fun as a photo album. "What was the day like outside on your birthday?" "Remember when we first saw a cardinal at our suet feeder?" "It snowed so much that day it came up to your belly button." Conversations and reminiscence create times of sharing and closeness. Recording events strengthens observational skills. All of it says to the child, "You and your life are special."

What To Do

1. Let the child draw a picture of himself or herself and the family.

2. Paste this picture on the cover of the diary.

3. Talk about how on each day many things happen to us. The things that happen are sometimes exciting. We might have a birthday, meet a new friend, eat something for the first time, do something new, like ride a tricycle, see a robin, hear water running in a stream, see an eagle, smell a skunk, go wading or swimming, have a new babysitter.

4. Encourage the children to make drawings of significant events as they occur.

5. Adults can label the drawings and give descriptions based on the children's narratives. Remember! Don't force this activity. On some days there won't be anything to write. Keep it fun for the kids.

6. Use photographs from home; for instance, pictures of a picnic, camping trip, birthday party, pets and so forth.

7. Let the children paste objects collected on walks on the pages.

8. Do texture rubbings and paste them in the diary.

9. After a period of time has passed, look back over the children's diaries. Have them tell about the significant happenings that took place.

Want To Do More?

Disassemble the diary book. Make a large mural of significant days of the month (a picture calendar). If a number of children participate, one page from each diary corresponding to the days of the month could be placed on the wall of a classroom. Don't forget to put each child's name on his/her contribution so that the books can be reassembled.

Make a nature scrapbook. Paste things in the scrapbook that children have collected on nature walks such as leaves, flowers, twigs, and seeds.

WATER CLOCKS

Things You Can Use

cans
water

Words You Can Use

time

Time is an elusive concept for children. But, measuring the passage of time is important. So let's find a way of keeping track of time by making a few timers. The students can be involved in a timing session with time measurers of their own.

What To Do

1. Collect five or six cans or plastic containers.
2. Clean the cans with soap and water.
3. Punch a small hole in the side of each can near the bottom. Try to make all the holes the same size.
4. Place a paper plug or cork in each hole.
5. Fill the cans with water.
6. Guess which one will empty first; which one last.
7. Release the plugs and compare their times. Which one did empty first? Why?

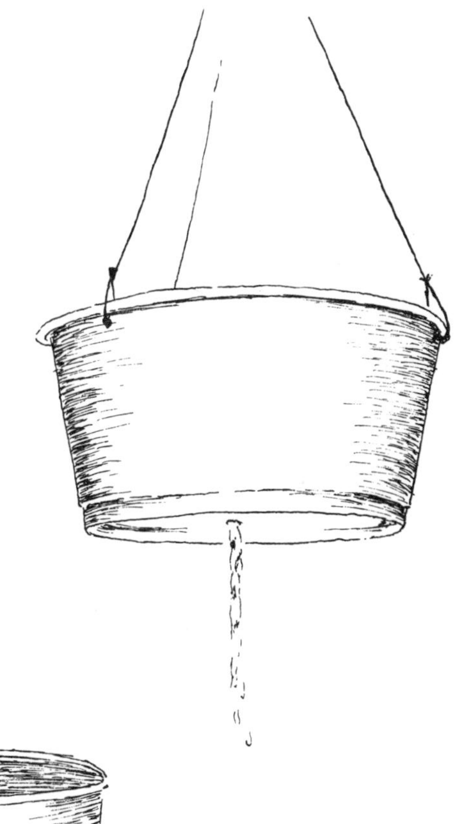

Want To Do More?

Choose common jobs or activities that can be timed with your water clocks. What is the effect of different sized holes? Compare the water clock to an hour glass or kitchen timer. Use a watch to calibrate the water clock; i.e. how much water flows out in one, two, three minutes?

HODGE PODGE

Some More Things We'd Like You to Know

Creating Super Duper Outdoor Lookers

Making the transition from play to these more structured activities can sometimes be difficult for young children. Something to help them focus on the task at hand is useful, especially if your outdoor exploring area is the same area you normally use for play. A simple technique we have found successful is the "Super Duper Outdoor Looker" tag. This tag is a circle of construction paper you hang around the child's neck and which has the child's name on one side and "Super Duper Outdoor Looker" on the other. Not complicated, but when it is presented with an air of seriousness mixed with excitement (after all this is an elite group!), it effectively indicates that SDOL time is different from playtime. An inexpensive magnifying glass, an "official" collecting bag, or the cardboard spyglass described in "A Spyglass Treasure Hunt" on page 33 can serve the same purpose. These props are not always needed however. Some of the most valuable learning occurs spontaneously when children and adults take time to notice what exists around them. Don't hesitate to spend a few minutes over the discovery of a caterpillar just because it's not part of a planned lesson.

Seating Arrangements

Seating arrangements can have a major influence on the success of your outdoor experiences. What you are doing will partly determine your seating, but in all cases, the children must be comfortable and be able to see. Arranging the children in a circle requires some time, but the circle allows greater interaction and teacher control. Obviously, for brief informal talks or information giving the standard "teacher facing the group" set up is most expedient. For times when a lot of sitting is required, carrying along a sit-upon (see page 100) is worth the trouble. It's a lot easier to be attentive with a comfortable behind.

Make a Snowflake

Making a snowflake is an exciting art if one is imitating a real snowflake. Because the snowflake's formation is controlled by the laws of nature, it is always a six sided figure. Have you ever tried to fold and cut a six sided figure? Here's how!

1. Select easily folded paper.
2. Draw a circle on the paper (a 2 lb coffee can is a nice size).
3. Cut out the circle.
4. Fold the circle shape in half.
5. Fold the half circle in thirds.
6. Make cuts anywhere on any side.
7. Unfold your six sided snowflake.

Seedbed in a Bag

Kids can make their own miniature plastic bag greenhouses and watch the processes that are usually mysterious underground phenomena.

Materials: ziploc bags, paper towels, seeds—beans, radish, peas, etc.

What To Do:

1. Place a damp paper towel in each bag.
2. Put some seeds into each bag - various kinds in separate rows or different kinds in different bags. Lock the bag.
3. Use masking tape to mark each bag with the gardener's name and the kind of seeds.
4. Arrange the bags on a shelf or tape to the bottom of window shades or blinds for observation. Place where heat is as close to 70 degrees as possible.

Watch to see which seeds sprout first and check daily on how they grow. There's no need to rewater the toweling if the bags are locked shut. The toweling will stay moist and the seeds will grow in about three weeks. Seedlings can then be transplanted to soil.

Make a Sit-Upon

A valuable piece of apparatus for exploring the outdoors is the sit-upon. This simply made cushion will keep your bottom dry when you sit down outside on damp days or days with heavy early morning dew.

1. Stacks of newspaper (one inch thickness of paper per child)
2. A plastic sack or piece of heavy plastic
3. Tape (masking or duct)
4. Magic markers (permanent)

What To Do:

1. Wrap the newspapers in the plastic sack.
2. Tape down loose ends.
3. With permanent markers have each child decorate a sit-upon.
4. Take an early morning walk. Now, when you sit upon the ground, you will all have dry bottoms.

Build a Plant Press

If you want to keep flowers and plants you collect on your trips, a plant press is a must. Here's how to construct a simple one.

1. Find 2 pieces of board of equal size. At least 40cm (16 in) long and 30cm (12 in) wide. These are your press supports, and are the same size as a folded newspaper.
2. Cut 20 pieces of cardboard the same size as your outside boards.
3. Collect newspapers.
4. Find ropes to hold everything together.
5. Alternate newspaper (8–10 pages per section) with the cardboard.
6. Put the stack of cardboard and newspaper between the two boards.
7. Place collected plants inside newspaper.
8. Tie *tightly* with rope.
9. Set in a warm, dry place, check in several days. Remove dried plants and store in a page of newspaper.

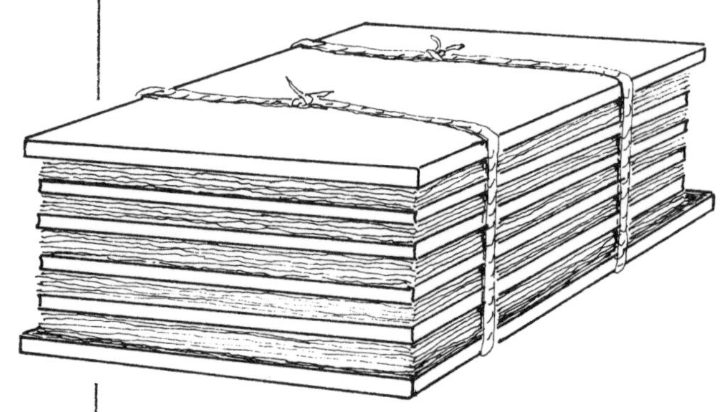

Poison Ivy - "Leaves Three, Leave It Be"

Poison ivy grows as a low shrub or vine. Its leaves are alternate, with three shiny green leaflets. It grows plentifully on flood plains, edges of woods, roadsides, fences, around buildings and along streams in Southern Canada, the United States and Mexico.

Try to avoid this plant on your walks. If someone does touch it, wash the affected area as soon as possible with hot soapy water. This will remove the oily sap that carries the skin irritant which is found in all parts of the plant. If a child develops a reaction to it, the itching, reddened skin and blister can be treated with soothing dressings of calamine lotion. Serious cases should be treated by a doctor.

Bites and Stings

There is always a chance that someone in the class will be stung by an insect. While on the trail, certain precautions should be taken to minimize the chance of being stung. You should:

1. Know how to identify insects that bite.
2. Avoid known insect nesting sites.
3. Don't wear perfume because bees and wasps are attracted to sweet scents.
4. Know your children's allergies.
5. Carry proper first aid treatment for those with allergy problems and carry first aid for just plain stings. If someone is stung during play time or on class walks, you will be prepared.

If stung, do the following:

1. Remove the stinger. It will look like a small splinter. Don't work the wound too much as you will force the venom into the skin.
2. Apply reaction prevention first aid, for instance:
    a. Place ice on the sting. This slows the reaction.
    b. An excellent pain and swelling reducer is meat tenderizer. This kitchen cooking aid is a powerful protein enzyme that breaks apart the protein from which venom is made. Use an unseasoned brand. When using the tenderizer, mix a small amount in your palm with water to make a paste. Place on sting and work in gently. Allow paste to set on sting. Moisten and remix occasionally.
3. Use a commercial sting and swelling prevention aid.
4. We have found the gel from an aloe vera plant to be a good remedy for stings also.

Ticks and Chiggers

Ticks and chiggers don't leave an immediate skin or allergic reaction but could effect a child in the evening or the next day after the field trip. Chiggers and ticks are spider-like animals that must burrow into the skin for a blood supply. Chiggers (red bugs) are tiny (almost too small to see) and will crawl on the body where the clothing fits lightly. Their activity can be retarded by spraying with an insect repellant or removed by scrubbing the body well with soap and water.

Ticks embed in the skin and are especially difficult to find in the hair. Each child should have a good hair brushing and have the scalp checked. Other places like the ears and arms should be checked. An insect repellant before leaving on the trip will stop many ticks from beginning the ride on a human body. Ticks and chiggers are both living members of our world whose survival depends on very predictable behavior. Knowing that behavior can make living with a distasteful creature an easier thing to do.

If a tick becomes embedded in a child's hair, a safe and easy way of removing it without medical care, is to cover the entire tick and its surrounding with a heavy coating of petroleum jelly. After 20 minutes the tick will be smothered, back out, and be easily pulled from its hold. Caution: Removing the tick forcibly without some aid could leave the tick's head and cause an infection.

Bird Feeders

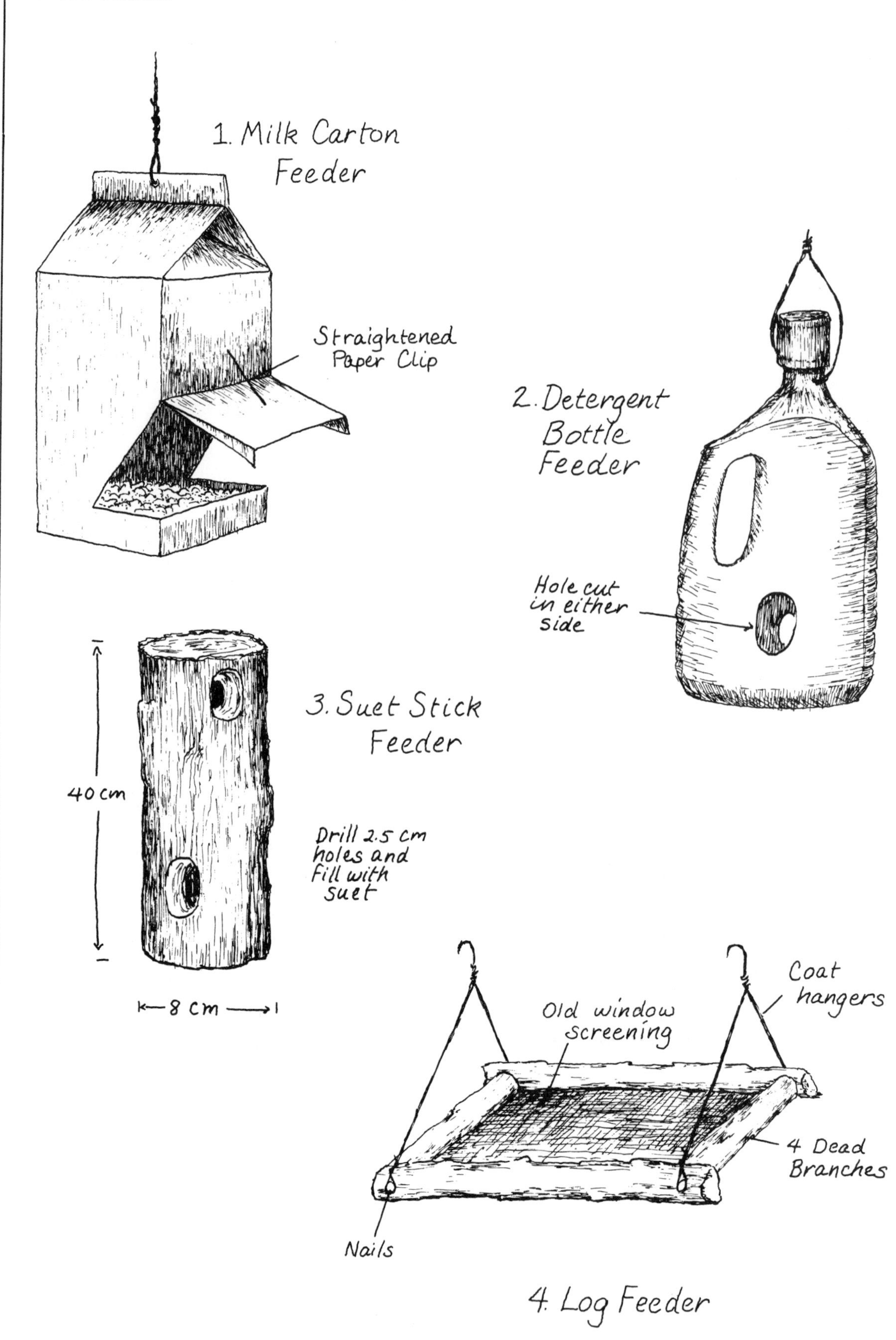

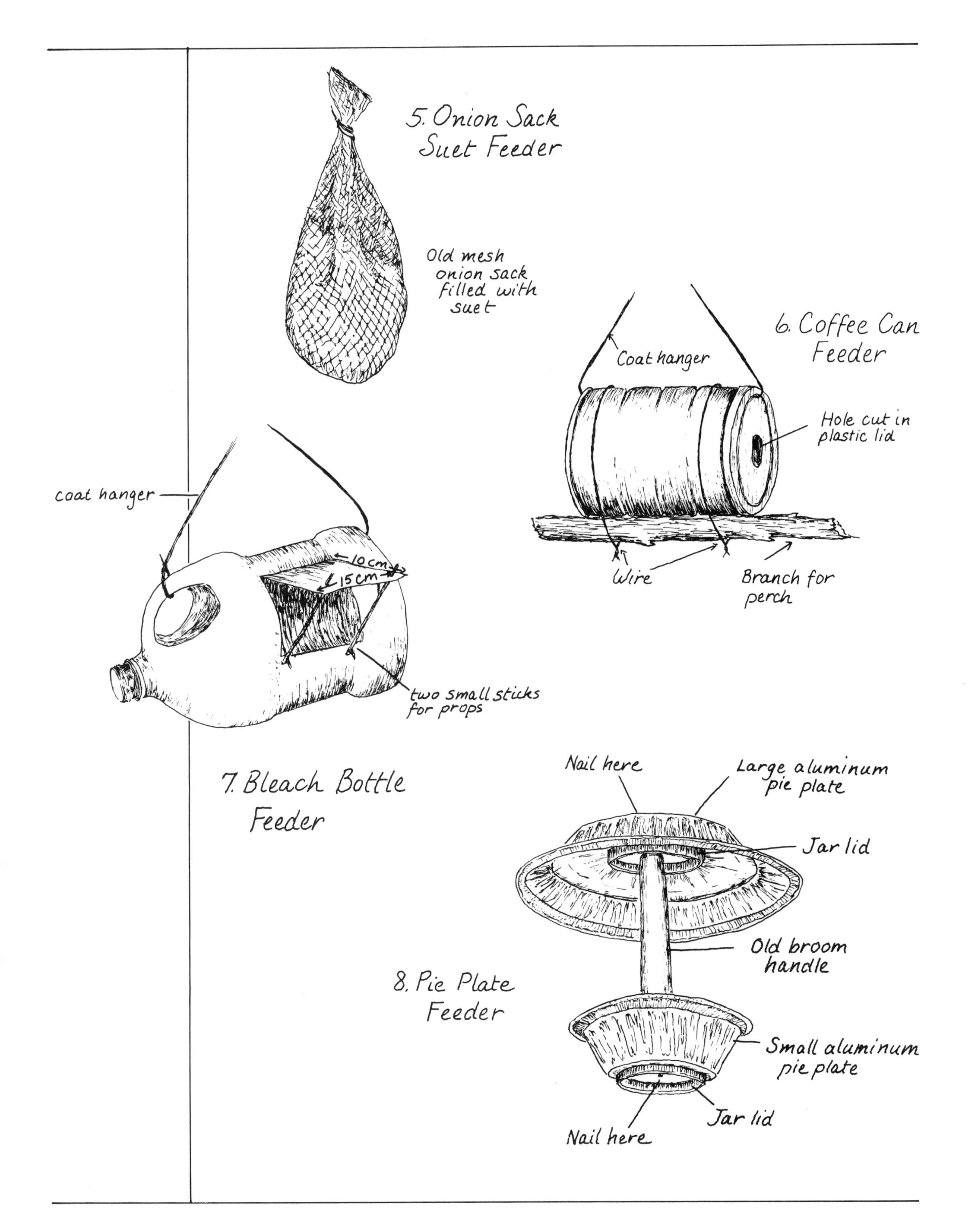
5. Onion Sack
Suet Feeder
Old mesh
onion sack
filled with
suet
6. Coffee Can
Feeder
Coat hanger
Hole cut in
plastic lid
Wire
Branch for
perch
coat hanger
10cm
15cm
two small sticks
for props
7. Bleach Bottle
Feeder
Nail here
Large aluminum
pie plate
Jar lid
Old broom
handle
8. Pie Plate
Feeder
Small aluminum
pie plate
Jar lid
Nail here

INDEX

Bubbles, Rainbows, and Worms:
Science Activities for Pre-School Children
by Sam Ed Brown

Bubbles, Rainbows, and Worms includes experiments with air, animals, the environment, plants, the senses, and water. Each experiment is complete with a learning objective, a list of materials, clear instructions, vocabulary words for language development, and an explanation for the teacher of the scientific principles behind the experiment. Dr. Brown is a former research chemist who changed careers to become a kindergarten teacher. He later became director of early childhood education for a major metropolitan school system, and now is Professor of Early Childhood Education at Texas Woman's University.

ISBN 0-87659-100-4, Paperback

Cooking With Kids
by Carolyn Ackerman

The emphasis is on active participation and nutrition. All of the recipes stress wholesome, natural ingredients. Artificial flavors, colors and preservatives are not used. Her suggestions to parents and teachers working with preschoolers are invaluable. For instance: "There may be times when a recipe doesn't turn out. Change it. Patch it up. Make it OK if you can. Remember, your ideas of what is acceptable is probably more stereotyped and narrow than your child's." Recipes include tempting treats such as Mountain Baked Potatoes, O-My-Pie, The Amazing Seed Sandwich, Moocumbers, and Blueberry Hill. The author is a research dietician, the mother of five children, and helped start Wheatsong, a whole grain bakery.

ISBN 0-88801-069-9, Spiral

Easy Woodstuff for Kids

by David Thompson

Very young children, even three-year-olds, can create beautiful things from wood. *Easy Woodstuff for Kids* will help children work with wood and give them a loving appreciation for trees and nature. All projects have a complete list of materials and tools, step-by-step instructions, and clear illustrations of what the project should look like every step of the way. The imaginative projects start with a simple stick name plaque. "The inventiveness extends beyond the design to the actual materials used—such as sticks, branches, and wood leftovers and scraps. The black-and-white graphics couldn't be better; neither the very detailed instruction and comments on each project. In short, a finely hewn book."—ALA Booklist.

ISBN 0-87659-101-2, Paperback

Available from bookstores and school supply stores or order directly from:

gryphon house inc.
P.O. Box 275
Mt. Rainier, MD 20712

One, Two, Buckle My Shoe: Math Activities for Young Children

by Sam Ed Brown

This clearly written, fully illustrated book contains 84 active, imaginative math activities for pre-school children. Each one creatively engages children's minds and bodies in learning fundamental math concepts. The activities are geared to the development needs of pre-school children, and use materials available in any home or classroom.

Content Areas:

Counting
One-to-one correspondence
Matching
Measurement
Shapes
Sequencing
Classification and Sets
Ordinal Numbers
Estimation and Future Planning
Numeral Recognition
Simple Addition and Subtraction

ISBN 0-87659-103-0, Paperback